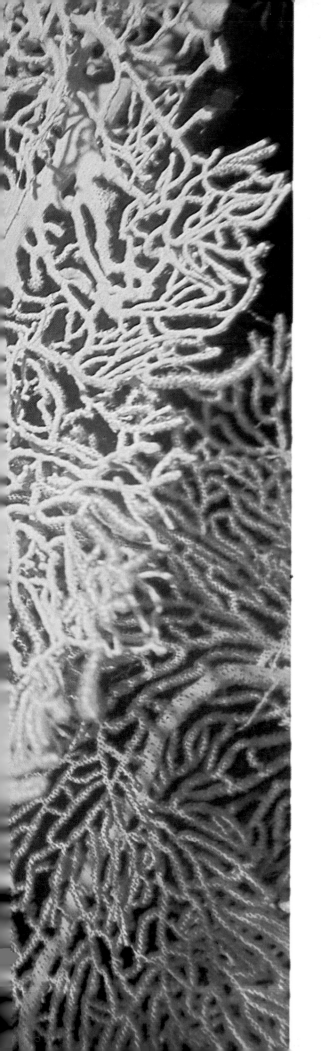

CORALS
of the
Great
Barrier Reef

Photography·Walter Deas
Text·Steve Domm

URE SMITH · SYDNEY

First published 1976 by
Ure Smith, Sydney
a division of Books for Pleasure Pty Ltd
176 South Creek Road, Dee Why West,
Australia 2099

Photography © Walter Deas 1976
Text © Steve Domm 1976

Designed in Australia by
Snape & Gallaher Graphics, Sydney
Typesetting by
ASA Typesetters, Sydney

Printed in Singapore by
Dainippon Tien Wah Printing (Pte) Ltd.,
977 Bukit Timah Road, Singapore 21.

National Library of Australia Cataloguing-in-
Publication data

Deas, Walter, 1933-, photographer.
  Corals of the Great Barrier Reef.

  Index.
  ISBN 0 7254 0265 2.

  1. Corals — Great Barrier Reef. I. Domm,
  Steven
  Bruce. II. Title.

593.609943

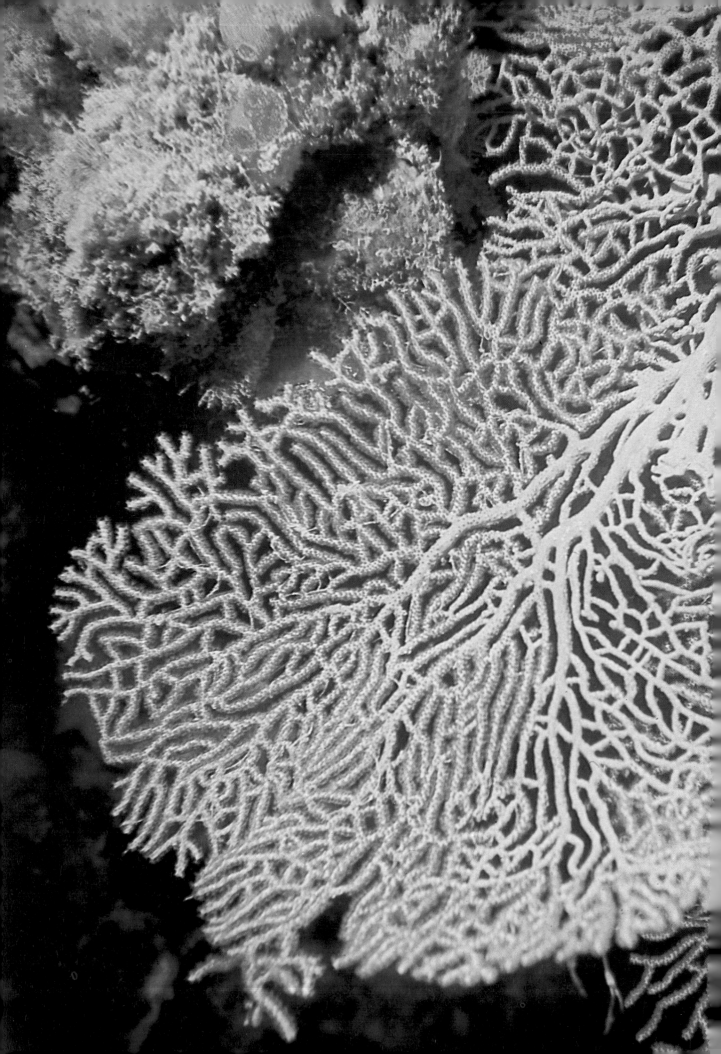

CORALS
of the
Great
Barrier Reef

Foreword

I welcome this book which presents corals as active living animals rather than as a series of intricately beautiful but inert limy skeletons as has been the case in the past. Too often corals are looked upon as attractive natural souvenirs—the kind that look fine on a room divider and have the advantage that they do not smell. Only the favoured few who have viewed them in all the glory of their living colour know their true beauty.

We frequently forget that they are sensitive and responsive living colonies whose polyps reach out voraciously to capture their food. The colonies take years to form. Like most other living creatures that make up the complex communities of coral reefs, they are subject to the same dangers and vicissitudes posed by the marine environment—adverse sea and weather conditions including cyclones, the most dangerous of all—and, above all, to the depredations and pollution that follow in the wake of so-called civilized man.

Scuba diving has provided a wonderful new way for naturalists to study coral reefs as living ecological communities. It is now more necessary than ever to be able to recognize and classify corals from the living colony underwater and not, as was done in the past, from examination of their cleaned, dead, limy skeletons. Underwater research and photography have made it feasible to record the appearance and habits of expanded corals and to marry these results with the old type of 'armchair' classification done in museums on dead material. It will no longer be necessary to maim or kill a colony in order to classify and record its ecology. This is a first step in providing a pictorial atlas or who's who of the living expanded corals of our Great Barrier Reef. As such it should be a boon not only to diver-naturalists but also be of tremendous interest to reef visitors.

Both authors have had considerable experience of the world of corals. Steve Domm has spent the last few years living on coral islands and conducting scientific experiments along Queensland's Great Barrier Reef. He has lived with them from day to day and, like the corals, 'sat out' cyclones and been on the spot to note the devastation and changes caused in the coral's world by raging seas and rains. He knows corals by day and by night, for some species are chiefly active at night. Walter Deas, now one of the world's most experienced underwater photographers, has worked in close collaboration with Domm in this project. Like his co-author, he measures his experience of living corals in terms of years rather than months.

Together they have produced an excellent introduction to coral ecology and a pictorial guide to the genera of Australia's Great Barrier Reef corals. It enables us to appreciate their diversity and beauty but, more important, it brings home to us the 'biological fragility' of this whole community. Hopefully it will swell the ranks of informed people who realize the need to conserve and protect the richest coral community in the world which, properly used and administered, could become one of Australia's greatest assets.

Elizabeth C. Pope, M.Sc.; C.M.Z.S.; F.R.Z.S.
Research Associate, Department of Marine
Invertebrates, the Australian Museum, Sydney

Acknowledgements

Dr David Barnes of the Australian Institute of Marine Science, Townsville, Qld, for his helpful comments on the manuscript

F.M. Bayer, Professor of Marine Science, University of Miami, USA, for assistance in identifying some of the soft corals

Jean Deas, diving buddy and underwater assistant, severest critic and loving wife of Walt Deas

Alison Domm, for many helpful comments and assistance both on the manuscript and with coral identifications

Fauna Preservation Society, London, who assisted with a grant towards the photographic work

Clarry Lawler, for the excellent line drawings

Dr W.G.H. Maxwell, for the information derived from his *Atlas of the Great Barrier Reef* and especially for his helpful comments on the manuscript

Lizard Island Research Station, for use of facilities by Walt and Jean Deas while photographing corals

Brian Parker, for the loan of equipment when all else failed

Pam Poulson, Heron Island Pty Ltd, for use of facilities during which time many of the underwater photographs for this book were obtained

Dr Michel Pichon of James Cook University, Townsville, Qld, for his helpful comments on the manuscript

Miss E. Pope, for the foreword and especially her very significant help with the manuscript

Carden Wallace, of the Queensland Museum, for checking the writer's coral identifications

Contents

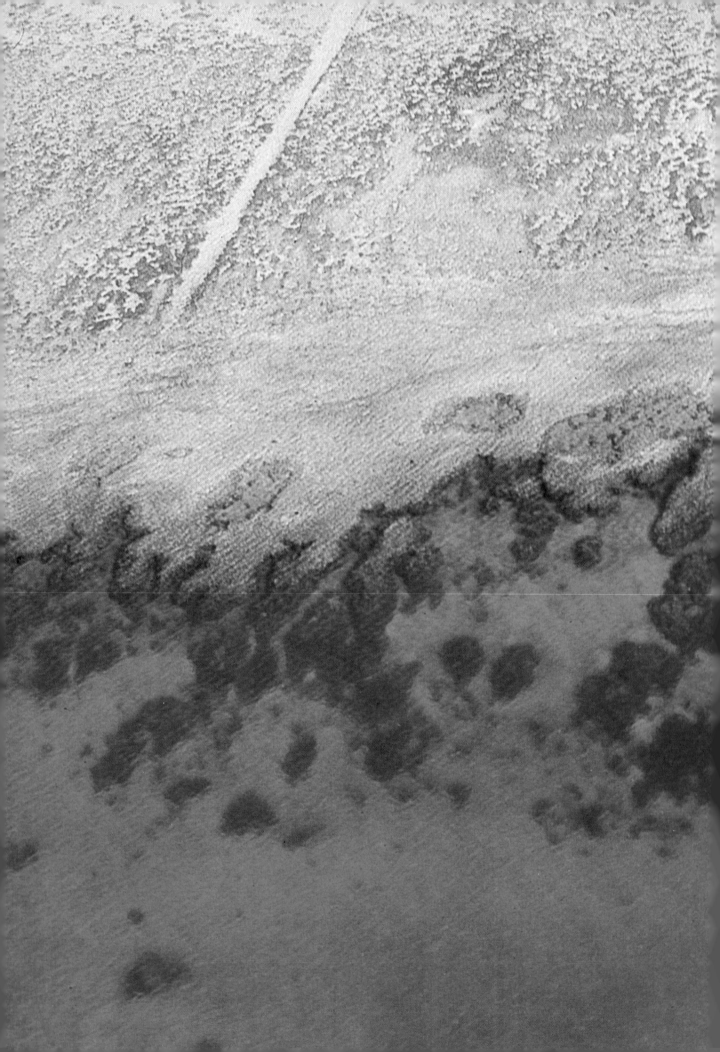

Coral Reef Aesthetics

To stand on a reef of the outer barrier and watch the surf crashing in, the air filled with thunder and sea mist, and to know that waves have been assaulting these same reefs for thousands of years, and yet they stand firm underfoot—this is the beginning of appreciation of the truly marvellous coral structures called reefs and the tiny architects responsible for them. For many years the writer and his wife have lived on islands of the Great Barrier Reef, the problems of coral reef research their constant companions: and yet they have never lost sight of the main reason for being here, which is a compelling love of natural beauty, solitude and the sea. Over a period of years the photographer and his wife have also made many visits to the Reef where they have endeavoured to record some of its wonders on film.

Coral reefs lie beneath the sea. Here in this silent world their beauty lies revealed. To swim amongst the corals and fishes, to be at home in their underwater world, is to enter a new dimension of awareness. The undersea landscape of living corals provides an exquisite setting for the other inhabitants of the reef. All possess beauty. In this tranquil world of subdued light, brilliantly coloured fishes with their often bizarre shapes add to the feeling of unreality. Perhaps it is this unique combination of beauty and strangeness that gives coral reefs their mystique.

The coral reef

Section 1 **Corals**

The Great Barrier Reef is not ours alone, but belongs to the future. Properly managed, it can provide an expanding income from tourism and fishing. However, the truly valuable aspect of the Reef cannot be measured in dollars and cents: it must be measured in enriched lives. Every year the Reef attracts Australians and visitors from overseas in increasing numbers. In an over-populated world natural areas of peace and beauty close to cities are becoming rare, and so Australia's Great Barrier Reef is an asset well worth preserving.

It is sincerely hoped that this book will assist towards a better understanding of the Great Barrier Reef. If intelligent and practical conservation policies are to be conceived and implemented, they must be the contribution of many persons representing both scientists and non-scientists alike. Conservation must be based on knowledge—and at the present time much of our knowledge of the Reef is tentative only. This book is written for the student of corals and coral reefs. It is assumed that the reader is familiar with fundamental principles of biology and geology. Scientific words have been used as little as possible to avoid confusing the non-scientific reader.

A diver ascends the coral reef wall

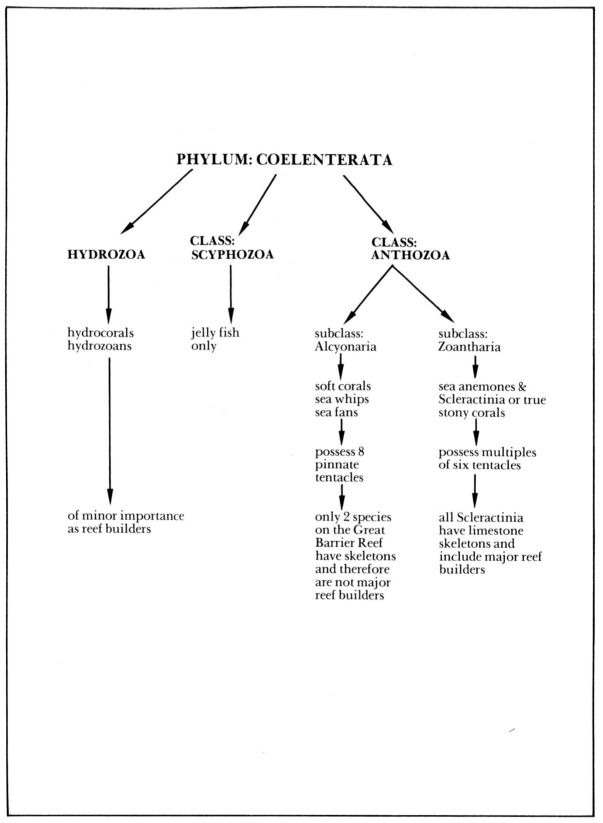

Fig. 1.

1. Coral Classification

Since this book is primarily concerned with the identification of corals, it is appropriate to introduce this section with a comment on the meaning of classification. The success of any study dealing with plants or animals depends upon the worker's ability to consistently recognize and name the organisms involved. To aid this, a world-wide system of naming plants and animals has evolved, based upon organizing them into groups which share similar physical characteristics. This is the meaning of classification. First there is the broad division into plants and animals. Each of these are then divided in *phyla*, groups that share very basic and fundamental structures. Each phylum is divided into different *classes*, classes into *orders*, orders into *families*, families into *genera*, and finally the different genera are again subdivided into the last category called the *species*. The species is the lowest and most important unit of classification, the members of the same species have the greatest resemblance to each other. Each organism is identified by two names: the first one, the generic name, establishes its relationship to other animals (or plants) in the classification; the second name, the species name, belongs to it alone.

Until the beginning of the eighteenth century corals were regarded as plants, presumably because of their sedentary nature and similarity to some plants on land. In 1744 they were first recognized as animals and in 1846 an effort was made to classify them. In 1857 the stony corals were recognized as a separate group. As collections became large enough to allow the comparison of a series of specimens, it became apparent that identifying coral species involved great difficulties. The growth form of corals, on which their identification so largely depended, was found to vary according to the particular habitat from which the coral was collected. In 1952, Cyril Crossland concludes in his great study of the corals of the Great Barrier Reef Expedition (Vol. VI), published in 1972 by the British Museum of Natural History, 'that the growth form varies so much that many so-called "species" are actually the mere products of the effect of different environmental influences, for example, shallowness of water, sedimentation or surf'.

The work of coral classification continues today, with workers from many countries throughout the world contributing. The problem of variation in species remains a very real one and is beginning to affect ecological studies. When working on the ecology of a reef, since the corals are often dominant animals, it is important to be able to identify all the species found. However, coral species are separated from each other by characteristics (usually in the skeleton) that are thought to be unique and non-variable. The problem lies in the fact that many 'species' change their growth form in response to their environment and it is difficult to distinguish real non-variable species from a series of different growth forms. At the present time the identification of many coral species is still uncertain and for this reason most corals will only be named to the generic level in this book.

It is difficult to define the word 'coral' because this term has been applied in popular usage to a heterogeneous group of sedentary marine animals often possessing different features. It generally includes three groups of animals: the stony corals (order Scleractinia), the soft corals (sea whips and sea fans) belonging to the subclass Octocorallia, and the hydrozoan corals (class Hydrozoa). See figure 1.

All Scleractinia secrete a hard limestone skeleton and the polyp (living animal) is characterized by having six or multiples of six tentacles surrounding its mouth. It has been estimated that there are 350 species of Scleractinia on the Great Barrier Reef belonging to approximately 60 genera. This book is mainly concerned with this group as they are the major reef builders and the dominant corals on most reefs.

The Scleractinia are divided into two main groups; the most numerous are the *hermatypic* corals which contain small plant cells (zooxanthellae) in their tissues and occur only in shallow and warm waters. The hermatypic corals are the reef-builders. The other group of Scleractinia are the *ahermatypic* corals which do not contain plant cells in their tissues and occur throughout the world in both shallow and very deep water. Ahermatypic corals can live in cold water but they are also present on coral reefs. The following genera occur on the Great Barrier Reef: *Heteropsammia, Dendrophyllia, Tubastrea* and others.

Corals belonging to the subclass Octocorallia are diverse in form but all have eight feathery tentacles on each polyp. The group referred to as the soft corals, or alcyonarians, are soft and leathery, and have a fleshy appearance. They lack a compact skeleton, their only form of support is derived from scattered needles (spicules) of limestone in their flesh. The sea fans and sea whips, or gorgonians, are mostly branching or tree-like in form and have a semi-rigid skeleton of a horny substance. There are two corals belonging to the Octocorallia that are unique because they secrete a limestone skeleton, and they may be confused with the Scleractinia. These are *Tubipora musica*, the so-called 'organ-pipe coral' having a red skeleton, and *Heliopora coerulea* having a blue skeleton.

The final group of corals comprise the hydrozoans, only two families of which secrete limestone skeletons; these are the Milleporidae and Stylasteridae. While these superficially resemble the stony corals they are structurally quite different. They are common on most reefs. There are other corals, such as the precious black coral, which are not described in this book as they are comparatively rare.

2. Coral Biology

Any study of the Great Barrier Reef must necessarily commence with the coral animal itself, for this exceedingly humble architect is essential in creating coral reefs throughout the tropical seas of the world. That an animal as small and simple as the coral polyp (soft part of the coral animal) can be a major contributor in the building of such durable, and often gigantic, structures as reefs must certainly rank as one of the biological wonders of the world.

a. Morphology

The coral polyp is very similar in shape and build to the common sea anemone, but it is generally smaller and it secretes a small cup-shaped structure of limestone (at its base) known as a *corallite*. The corallite provides support and protection for the soft-bodied polyp which is able to retract into the corallite when not feeding. The polyp is a simple structure having a crown of tentacles surrounding a central 'mouth' which opens into the hollow cylindrical body. While coral colonies may form a great many shapes and sizes due to different skeletal patterns, the coral polyp, which is the fundamental unit, remains the same. It is within this hollow body and the cells lining it that digestion and sexual reproduction take place. The tentacles surrounding the mouth are armed with stinging cells that enable the coral to capture and kill (or immobilize) the small floating animals (zooplankton) which form a major part of its diet.

The extended polyps of *Tubastrea* sp. illustrate the general shape of a coral polyp. Photograph taken at night. (Heron Island)

Some corals occur as single solitary polyps, but most grow in the form of colonies comprising numerous individuals. The skeleton of such a colony comprising many corallites is referred to as a *corallum*. In a colony the polyps are interconnected by tissues that lie as a thin skin over the non-living skeleton. Each colony continually secretes limestone and, as a result, the colony expands outwards and upwards.

Colonies of stony corals occur in many different shapes and the intricate sculpturing of some coral skeletons is very beautiful, while others are so fragile it is difficult to understand how they survive. The most common growth forms are the following:

branching colonies that resemble trees or shrubs,
round massive colonies that resemble boulders,
plate or 'table' corals that grow as broad circular upward-facing colonies usually supported by a central column,
corals that occur as thin sheets sometimes called 'leaf' coral,
encrusting colonies that grow closely over the reef surface,
free-living corals that occur unattached on the sea bottom,
solitary corals that consist of a single polyp—often relatively large.

While it is the skeleton of a coral that gives it such an attractive shape it is the living animal's tissues that provide the colour. Remove the living tissue and, with few exceptions, a white skeleton remains. Contrary to tourist brochures, corals are not always colourful. The majority are subdued shades of brown. It is necessary to search for the more vivid specimens. These however can be exquisitely beautiful in blues, mauves, greens, oranges and, sometimes, red.

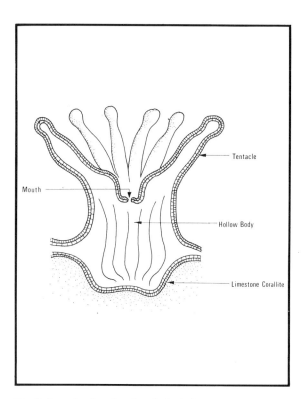

Fig. 2. A coral polyp showing the basic features

This tropical anemone is a close relative of coral and harbours an anemonefish, *Amphiprion melanopus*, which lives in a symbiotic relationship with the anemone. (Heron Island)

b. Feeding

Corals are carnivorous animals but they also obtain some nourishment from the plant cells (zooxanthellae) embedded in their tissues. Food in the form of zooplankton is captured by the tentacles. A basic feature of all corals, and one shared by other animals of the phylum Coelenterata, is the presence of stinging cells on the tentacles. The stinging cells contain nematocysts which discharge like small hypodermic needles into any small animal that touches the tentacles. A poison is injected which immobilizes or kills the animal, which is then transferred by the tentacle to the mouth and then to the digestive cavity. Most coral stings are not sufficiently powerful to be felt by humans, however, certain hydrozoan corals *(Millepora)* can cause mild stinging and are called 'fire corals'.

Most corals feed at night and at this time their polyps are fully extended in search of food. During the daytime they are retracted within their corallites. Because of this, many visitors to the Reef leave without having seen expanded coral polyps, most of which resemble small flowers. The night-feeding pattern of corals probably occurs because zooplankton on reefs are most active at night.

Recent scientific research has begun to explore the question of how much nutrition the coral obtains from zooplankton on the one hand and from its zooxanthellae on the other. Some scientists believe that corals obtain most, if not all, of their nutriment from their zooxanthellae, while others seem to think zooxanthellae play only a minor role in providing basic nutrition.

c. Reproduction and Growth

Corals reproduce by two methods. The first, asexual reproduction or 'budding', results in the growth of colonies by the subdivision of existing polyps; the colony expands as a result of budding by growth in various directions depending upon its growth form.

The second, sexual reproduction, results in the formation of very small free-swimming larvae called *planulae*, which swim away from the parent colony, settle and begin a new colony. In some colonies the sexes are separate, i.e. the colony is either male or female, while in others both sexes occur in the same colony. Generally, fertilization of the egg takes place within the body of a female polyp when sperm released from a male polyp of the same species enters the 'mouth' and makes contact with the egg. The small multicellular (many celled) planula develops inside the polyp and is later released to begin a free-swimming life. The release of planulae is seasonal in some corals (generally summer months) and in others, e.g. *Pocillopora damicornis,* appears to be controlled by the phases of the moon throughout the year. The planula swims slowly through the water, using its hair-like projections for locomotion. After a variable period, if it survives predation and other adverse circumstances, it will settle and grow into a polyp. First it begins to secrete a base of limestone and then it develops tentacles and a mouth and soon it is a functional polyp, i.e. it is able to capture food. It then begins to bud off a small additional polyp from one side, then another and so on, until a small colony has formed. During this early stage mortality is often very high but the large numbers of planulae released compensate for this. There is often considerable competition for space on a reef and the small colony may be eliminated by adjacent corals or other organisms which restrict the light or which may even eat it.

It is not known how long corals live and it undoubtedly varies from species to species. Large brain corals have been recorded in one location for many years and life expectancy in corals to a large extent is dependent upon environment. Favourable environments that are stable over many years would be conducive to coral colonies attaining a great age.

d. The Coral-Zooxanthellae Relationship

Corals would be rather unremarkable animals if it were not for their ability to secrete calcium carbonate (limestone) in so efficient a manner. Indeed, all modern coral reefs owe their existence to the presence of zooxanthellae within the tissues of hermatypic corals, since it is these small plant cells that are largely responsible for the remarkable powers of skeleton formation in corals. These skeletons later become a major component of a reef. A knowledge of the relationship between the coral and its zooxanthellae is therefore of fundamental importance.

The presence of zooxanthellae within the tissues of corals was discovered many years ago. However it is only recently that the details of this working partnership have become clear. The relationship between the coral and the zooxanthellae is *symbiotic,* which means it is of mutual benefit to both partners. However, it is not an essential relationship in that the coral can survive without the zooxanthellae. The zooxanthellae have recently been identified as the vegetative non-motile phase of a *dinoflagellate.* Dinoflagellates are unicellular yellow-brown algae which have a characteristic pair of flagellae used in locomotion during their motile phase of existence. The simplest method of understanding the partnership is to enumerate the benefits to each organism. First, let us consider the ways in which the zooxanthellae benefit, recognizing that they are plants requiring sunlight, carbon dioxide and certain simple inorganic substances to survive and grow.

1. The zooxanthellae receive protection within the tissues of the coral.
2. The zooxanthellae use the carbon dioxide produced by coral respiration to make carbohydrates by photosynthesis.
3. The zooxanthellae absorb compounds containing nitrogen and phosphorous, which are the excretory products of the coral, and make them into new proteins and other useful substances. The zooxanthellae can obtain greater concentrations of these vital substances from within the coral's tissues than would be available if they were free living in the sea.

Now, considering the ways in which the coral benefits from the partnership, we find that it is mainly through an enhancement of metabolic efficiency, some nutritional gain and through an increased ability to lay down or secrete limestone and therefore build skeletons. In more detail, we find the presence of zooxanthellae help corals as follows.

1. The removal of waste products carbon dioxide, nitrogen and phosphorous by zooxanthellae greatly enhances or speeds up the coral's metabolic efficiency—possibly because the simple polyp lacks the specialized organs that carry out these processes in the higher animals.
2. The production of oxygen by the zooxanthellae during photosynthesis may be of some value to the corals.
3. Recently it was discovered that the zooxanthellae make an important contribution to the nutrition of the host coral. It was found that when the zooxanthellae utilized carbon dioxide, they released substantial quantities of soluble carbohydrates which were metabolized by the coral. In one species of stony coral, from 30 to 50 per cent of the total carbon compounds produced by the zooxanthellae were incorporated into the coral's tissues. It is not known exactly what proportion of the coral's dietary needs are satisfied in this way and most corals probably supplement their diet by capturing zooplankton.
4. Perhaps the most significant benefit to the coral is the speeding up of the process of skeleton formation. The exact mechanism by which the zooxanthellae enhance calcification is not as yet well understood and involves rather complex chemical reasoning which will not be pursued further in this book.

We have now dealt with the roles played by both the coral and the zooxanthellae in their mutually beneficial relationship. The coral-plant relationship is a small but complete ecosystem within itself. The use of carbon dioxide, oxygen, phosphorous and nitrogen is cyclical. As carbon dioxide, phosphorous and nitrogen are excreted

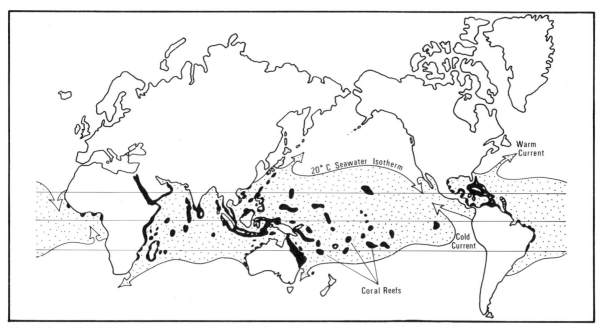

Fig. 3. The coral reef regions of the world are found within the 20°C (80°F) seawater isotherm and are therefore largely confined to the tropics

by the coral, they are absorbed by the zooxanthellae which convert them to useful biological substances once again. These are in turn used by the coral—oxygen for respiration; proteins, lipids and carbohydrates for incorporation into the coral's tissues. The coral exhibits greater metabolic efficiency because the zooxanthellae play the role of 'organs of excretion'. Finally, the zooxanthellae greatly speed up the coral's ability to form limestone skeletons. All these factors, which so greatly benefit the coral, indirectly benefit the entire coral reef community since corals are of major importance in reef formation and maintenance.

e. Ecology

The geographic (world-wide) distribution of hermatypic corals and the reefs they help to form is controlled by a number of environmental requirements. The most important of these is temperature. Hermatypic corals will only flourish in warm oceanic water between 20°C (68°F) and 30°C (86°F) and this factor confines reef-building corals mainly to the tropics. Where hermatypic corals exist in cooler waters they generally do not form reefs and their breeding is probably impaired. The Great Barrier Reef lies within the tropics except for the southern end (Bunker Group of reefs and islands) which is just south of

Crustaceans such as these bore into coral skeletons and do minor damage. (Heron Island)

Above: The tube worm *Spirobranchus giganteus* lives within the skeleton of living corals, and is most common in species of *Porites*. (Heron Island)

Below: Small female crabs, *Hapalocarcinus marsupialis*, induce the formation of galls or chambers in certain corals. They live permanently trapped within these, thus having a unique relationship with the living coral. Small apertures remain in the gall through which the female filters food and by way of which the tiny male crab can gain access to the female. Illustrated is the coral *Seriatopora* showing several galls. (Heron Island)

Right: Corals provide 'homes' for many animals on a reef, especially the fishes. (Heron Island)

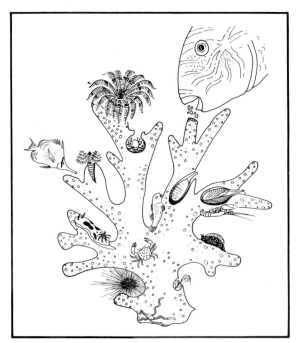

Fig. 4. Many animals use living corals for shelter, but few feed directly on them; butterflyfish and parrot fish are two exceptions

Left: Many species of butterflyfish feed on coral polyps, but do minor damage. Species illustrated is *Chaetodon trifascialis*. (Wistari Reef)

the tropic of Capricorn. Here, fewer coral species are found than further north.

Another important factor governing coral distribution is light, a requirement of the associated zooxanthellae (plant cells) rather than the coral animal itself as all plants require light. Light penetration limits the depth at which hermatypic corals will grow. Flourishing coral growths are only found in shallow water between approximately 2 and 13 metres (6-43 ft) in depth with decreasing growth down to about 60 metres (197 ft). The coral's need for shallow clear water explains why reefs are generally found in shallow areas on continental shelves (shelf reefs), or associated with islands (fringing reefs), or on sunken mountain tops (atolls).

Corals have a limited tolerance to sediment, salinity changes, and exposure to air. They also require adequate water movement to supply the food they need. These factors all influence the distribution of reefs and corals.

Let us now consider the distribution of corals on an individual reef. Some areas support flourishing coral growths and others are impoverished by comparison. A reef does not provide a homogeneous environment, e.g. the reef top or reef flat has a totally different environment from the reef slope and the reef slope differs according to whether it is protected from the prevailing wind and waves or exposed to them. Each of these areas experiences variations in depth, wave action,

water movement, temperature, salinity, sedimentation and other factors more difficult to measure. All these factors influence coral distribution and abundance. Certain coral species favour a broad area or zone on a reef and are biologically adapted to the conditions found in that particular area.

At the present time little work has been done on how corals growing on a reef affect each other. There is much competition between organisms for space and a high degree of crowding occurs. The faster growing corals will often tend to eliminate the slower growing ones by preventing sunlight from reaching adjacent but smaller or slower growing colonies. It is believed that some corals (probably having stronger nematocysts) are able to kill and digest neighbouring corals and thereby become dominant.

Corals are the most important group of animals on a reef contributing to general habitat diversity. Coral reefs are characterized by having an extremely irregular surface and topography, with innumerable holes, caves, overhangs and similar features. Generally, a more complex environment will support more animals of different sorts than a simple one. Large numbers of fishes, invertebrates and other organisms find shelter amongst the varied forms of the living corals.

Strangely enough, considering the dominance of corals, relatively few reef animals feed directly on corals. This may be due to the numerous nematocysts on the tentacles. Some fishes (e.g. parrot fishes and butterflyfishes), along with several species of marine snails, eat small amounts of coral polyps, however, this damage is seldom significant.

The one exception to the above remarks is, of course, the Crown-of-Thorns starfish, *Acanthaster planci*. It is only since the early 1960s that marine scientists have discovered this starfish in large numbers killing and eating the stony corals on certain reefs. The problem of *Acanthaster* infestation is not confined solely to the Great Barrier Reef, but is an Indo-Pacific phenomenon causing damage throughout this area. That a problem exists is now beyond doubt, but its magnitude and long-term significance as a threat to the entire Reef is not understood.

When it is realized that the stony corals are key species in the overall reef ecosystem, it can be appreciated that the destruction of all or most of the living corals on a reef would have a drastic and immediate effect on the entire reef community. The Crown-of-Thorns starfish only digests the living tissues of a coral and not its skeleton and therefore the actual structure of the reef remains. However, the process of coral reef recolonization is slow and would initially result in a veneer of different plants and animals with a changed ecology.

No satisfactory explanation has been found for the proliferation of this previously uncommon starfish. Some people claim it is a natural cycle, while others think it is caused in some way by

The Crown-of-Thorns starfish, *Acanthaster planci*, is causing serious damage to some reefs of the Great Barrier Reef.
Left: A large starfish on an *Acropora* coral. The white region has been preyed upon. (Lizard Island)
Above: A close-up showing the tube feet and spines. (One Tree Island)

Below: Jean Deas inspects a Crown-of-Thorns starfish in the lagoon at One Tree Island.

man's interference with the Reef. There is also a controversy about what measures should be taken in order to control the Crown-of-Thorns.

Like many other biological problems, more detailed information is required about this starfish's biology and life history, plus its long-term effect on coral reefs before intelligent action can be taken.

3. Corals of the Great Barrier Reef

Since it is very difficult to describe corals without resorting to a complex scientific terminology, the written descriptions of corals in this book are kept to an absolute minimum. It is anticipated that most coral identifications will be made by referring to the photographs and this is the best method available to the amateur and to zoologists who are not coral specialists.

Since corals are variable in colour and form, these characteristics must be treated with caution. The colours displayed by most coral species can

A coral reef is the home of the giant clam, *Tridacna gigas*, and its smaller relatives, *T. maxima* and *T. crocea*.

Left: Jean Deas with a giant clam, *T. gigas* in the lagoon at Lizard Island.

Below: The small clam, *T. maxima* is noted for the splendour and colour of its fleshy mantle. (Heron Island)

Top right: The clam *T. crocea* is deeply embedded in the algae-covered reef. (Lizard Island)

Bottom right: The characteristic appearance of limestone secreting algae which are so important in reef formation. (Heron Island)

The characteristic appearance of limestone secreting algae which are so important in reef formation. (Lizard Island)

vary from one reef to another (and are more intense and varied in northern waters) or even from different areas across the same reef. Also the time of year may influence colouration. The gross form of many corals is variable and can change in different environments. Size is relative to maturity and, again, environment. Thus the reader is urged to treat statements about colour, form and size not as absolute criteria, but as commonly occurring conditions and to expect variation.

Generally the identification of corals is based upon detailed characteristics of the corallites, their relationship to the corallum, and the corallum or gross form itself. However, features that may be apparent in the skeleton, especially with a hand lens or microscope, may not be apparent in the living coral out on the reef. The photographs in this book represent what the author considers to be an average or common growth form for each genus taken alive on the Reef.

Scleractinia

Family: THAMNASTERIIDAE
Genus: *Psammocora*
Distribution: Reef flat
Characteristics: Uncommon. A rather obscure coral having a smooth surface with very small, indistinct corallites. The two most common species are one with low, rounded branches and one which grows as small, rounded colonies. The colonies average 10 centimetres (4 in.) across. Common colours are greyish and burgundy.
Lizard Island

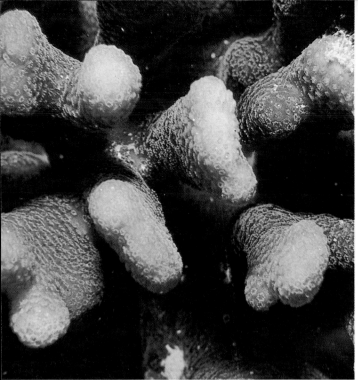

Family: POCILLOPORIDAE
Genus: *Stylophora*
Distribution: Reef flat, reef slopes and lagoon
Characteristics: Common. *Stylophora* occurs as a clump of rather thick branches having a slightly rough feel to their surface when touched. An average colony might be 30 centimetres (12 in.) across, and colours can be purple, blue, brown and pink.
Below: Heron Island; Left: Close-up, Heron Island

29

Family: POCILLOPORIDAE
Genus: *Seriatopora*
Common name: Needle Coral
Distribution: Reef flat, reef slopes and lagoon
Characteristics: Common. This fragile coral occurs as rather dense 'thickets' of slender, interlocking branches. Colonies vary from 10-50 centimetres (4-20 in.) across. Common colours are pink, brown, green and grey, often with whitish tips.
Below: Lizard Island; Top right: Heron Island; Right: Kenn Reef

Family: POCILLOPORIDAE
Genus: *Pocillopora*
Common name: Brown Coral
Distribution: Reef flats, slopes and lagoon
Characteristics: Common. The most common species, *Pocillopora damicornis,* grows as a dense bushy cluster of smooth branches. Another species, more common to the north, grows in large colonies of upright lobes, the surface of which is covered by numerous rounded projections. Colours are brown, pink or green and colonies range from 10-30 centimetres (4-12 in.) across.
Bottom: Heron Island; Below: Heron Island; Left: Lizard Island

Family: ACROPORIDAE
Genus: *Acropora*
Common name: Staghorn Coral
Distribution: Sheltered locations, including reef slopes, lagoon and reef flat
Characteristics: Common. This prolific genus comprises many species and is the dominant coral on most reefs. The growth form varies markedly from species to species, but four kinds are particularly typical: branching or staghorn colonies, bushy colonies, plate-like formations, and low colonies that may be encrusting in form. *Acropora* species are characterized by the presence of many small cylindrical corallites that stand out from the corallum. The colour is variable, sometimes brilliant, and it can be blue, purple, pink, yellow, cream, orange, green or brown. Colony size varies from a few centimetres to several metres across.
Right: Lizard Island; Below: Heron Island; Top page 31: Sykes Reef; Bottom page 32: Close-up, Heron Island

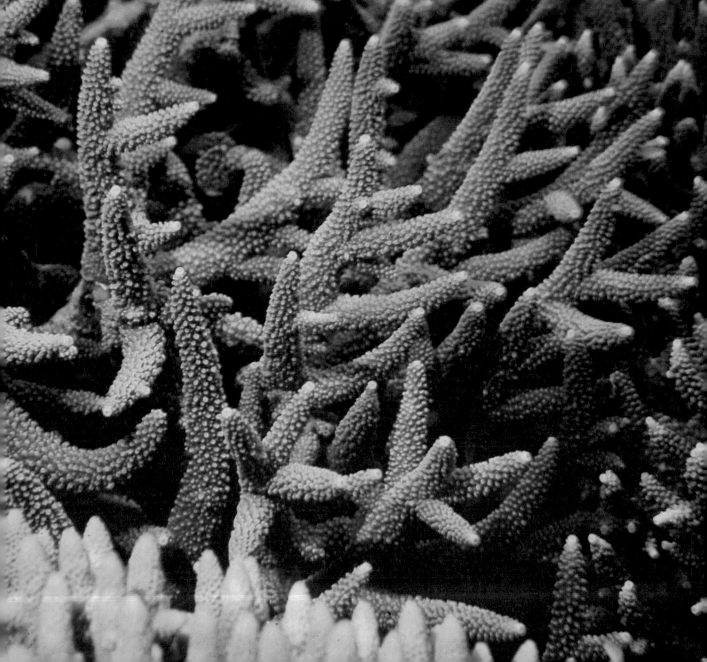

Family: ACROPORIDAE
Genus: *Acropora*
Common name: Staghorn Coral
Characteristics: These corals represent a slightly different growth form of the more typical staghorn, having a more bushy appearance.
Below: Palfrey Island; Right: Lizard Island

Family: ACROPORIDAE
Genus: *Acropora*
Common name: Bush Coral
Distribution: Entire reef surface where conditions are favourable
Bottom: Lizard Island; Left: Palfrey Island; Below: Heron Island

Family: ACROPORIDAE
Genus: *Acropora*
Common name: Plate Coral
Distribution: Most common in semi-protected areas of reef slopes
Bottom: Lizard Island; Right: Heron Island; Below: Heron Island

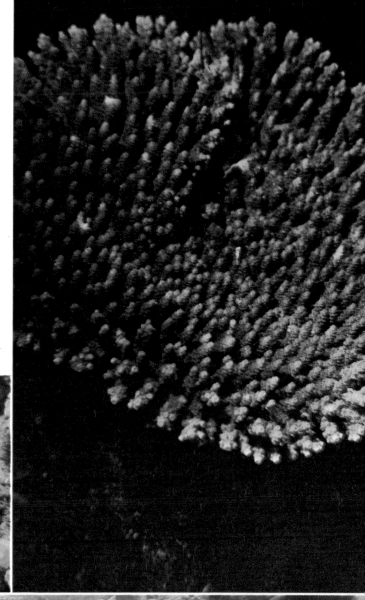

Family: ACROPORIDAE
Genus: *Acropora*
Common name: Knobbly Coral
Distribution: Reef flat and shallow reef slopes
Below: Lizard Island; Left: South Island

Family: ACROPORIDAE
Genus: *Astreopora*
Distribution: Reef flat and upper reef slopes
Characteristics: Uncommon. This coral usually occurs as rounded colonies, varying in size from 10 centimetres (4 in.) to 1 metre (3 ft) across. All are characterized by slightly protuberant corallites which are larger than those of *Montipora* and have an 'empty' appearance. Common colours are pink, green and pale blue.
Below: Lizard Island; Bottom: Wistari Reef; Right: Heron Island

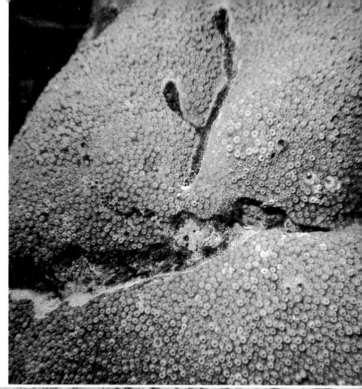

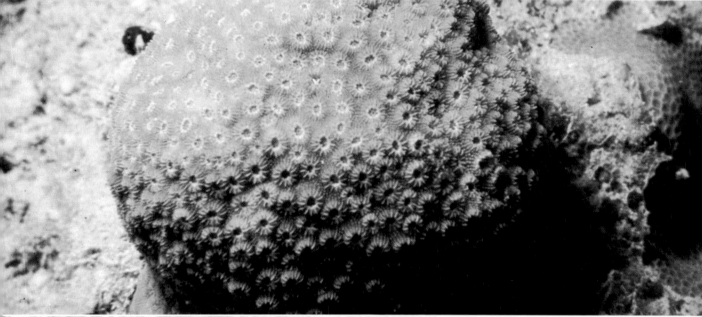

Family: ACROPORIDAE
Genus: *Montipora*
Common name: Leaf Coral
Distribution: Reef flat, slopes and lagoon
Characteristics: Common. This coral, of which there are many species, is usually found growing as leafy plates or as an irregular encrustation. There are also some branching species. *Montipora* has a rough surface due to the many small corallites and projections. The colour is variable, with browns, pink, red and green predominating. The size of the colonies varies from a few centimetres across to 2 metres (6 ft) or more.

Below: Heron Island; Bottom: Heron Island; Left: Lizard Island

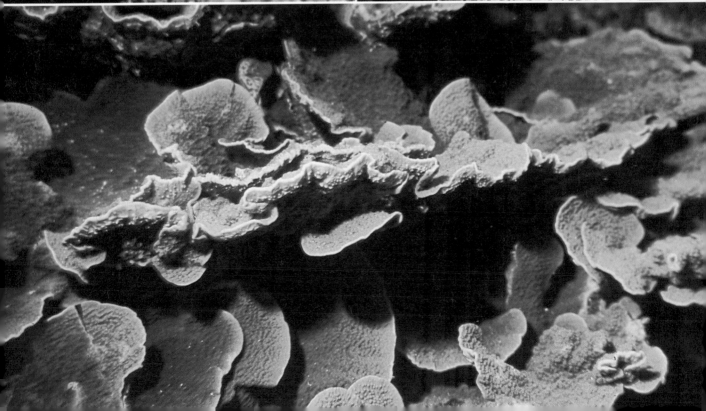

Family: AGARICIIDAE
Genus: *Pavona*
Distribution: Reef flat and shallow reef slopes
Characteristics: Uncommon. Can be found in various growth forms and comprises several species. The identification of this coral can be difficult. However, a characteristic feature of many species is the radiating pattern of the major septa which radiate outwards from the centre of each corallite, like spokes from a wheel. The size is highly variable and the colours are usually brown, grey or green.

Below: Heron Island; Top right: Close-up, Lizard Island;
Right: Close-up, Lizard Island

Family: AGARICIIDAE.
Genus: *Pachyseris*
Distribution: Reef slopes
Characteristics: Uncommon. There are two species of this coral, each having a distinctive growth form. However both are characterized by the presence of ridges, often concentric, on their surface. Generally, size would be 20 centimetres (8 in.) across and the most common colour is brown.
Below: Lizard Island; Left: Lizard Island

41

Family: FUNGIIDAE
Genus: *Fungia*
Common name: Mushroom Coral
Distribution: Reef flat, slopes and lagoon
Characteristics: Common. Each coral, a solitary polyp of large size is found lying free on the bottom, except in the early stages of development, when it is attached to the substrate. There are several species of *Fungia* varying in shape from round to oval; with one exception all have their tentacles retracted by day. *Fungia actiniformis* has its tentacles out all of the time and resembles an anemone. Colour is normally brown and sizes vary from 5-20 centimetres (2-8 in.) across.
Various Fungia *spp; Heron Island;*

Above: Heron Island

Above: A juvenile, Lizard Island

Above: Shows a number of juvenile Fungia *in various stages of development, One Tree Island*
Above right: Fungia actiniformis, *Heron Island*

Below: Close-up of the polyps of Fungia actiniformis, *Heron Island*

43

Family: FUNGIIDAE
Genus: *Herpolitha*
Common name: Slipper Coral
Distribution: Reef slopes and lagoon
Characteristics: Uncommon. The elongated shape and central groove of the Slipper Coral make it easy to recognize. It is a colonial fungiid, lying free on the bottom. The colour is brown and its average length is 25 centimetres (10 in.).
Heron Island

Family: FUNGIIDAE
Genus: *Polyphyllia*
Common name: Slipper Coral
Distribution: Reef slopes and lagoon
Characteristics: Uncommon. This coral has a superficial resemblance to *Herpolitha*, but lacks the central groove and is more dome shaped. It is colonial and free living. The colour is brown and an average length would be 25 centimetres (10 in.).
Below: Heron Island; Left: Close-up, Heron Island

Family: FUNGIIDAE
Genus: *Parahalomitra*
Common name: Basket Coral
Distribution: Reef slopes and lagoon
Characteristics: Uncommon. This large colonial fungiid lies free on the bottom, and grows in the shape of a dome. The colour is brown and an average diameter would be 25 centimetres (10 in.).
Below: Tryon Island; Right: Heron Island

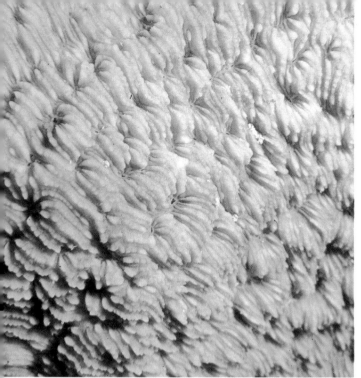

Family: FUNGIIDAE
Genus: *Podabacia*
Distribution: Lagoon and sheltered reef slopes
Characteristics: Uncommon. This is the only member of the Fungiidae that, as a mature colony, grows attached to the bottom. It grows as thin sheets or in a vase-like shape and has an irregular surface resembling the other fungiids. An average colony might be 30 centimetres (12 in.) across and the colours are commonly browns.
Below: Lizard Island, photo: Steve Domm; Left: Close-up, Lizard Island

Family: FUNGIIDAE
Genus: *Cycloseris*
Distribution: Sandy bottom adjacent to reefs at depths below 10 metres (33 ft)
Characteristics: Uncommon. This small coral, like many fungiids, is a single polyp and occurs free on the bottom. It is usually green or grey in colour and averages 2 centimetres in diameter. *Cycloseris* is the most primitive of the Fungiidae. The earliest forms occurred in the Middle Cretaceous period.
Lizard Island

Family: FUNGIIDAE
Genus: *Diaseris*
Common name: Acrobatic Coral
Distribution: Sandy bottom adjacent to reefs below a depth of 10 metres (33 ft)
Characteristics: Uncommon. This is the smallest of the Fungiidae. It is usually only 1 centimetre in diameter and rather fragile, breaking easily into 2 or 3 fragments along well-defined fracture lines. Like most fungiids, *Diaseris* lies free but it has exceptional powers of locomotion and is able to move along the bottom, turn itself over and can even 'escape' from a small dish.
Lizard Island

Family: PORITIDAE
Genus: *Goniopora*
Distribution: Reef flat, slopes and lagoon
Characteristics: Common. This coral has its polyps out by day and could be mistaken for a soft coral. *Goniopora* is usually grey, brown or green in colour, and grows as round colonies having an average diameter of 20 centimetres (8 in.) but it can also form colonies over 1 metre (3 ft) in diameter.
Top right: Heron Island; Right: Polyps emerging, Heron Island, photo: Jean Deas; Below: Close-up of polyps, Heron Island

Family: PORITIDAE
Genus: *Porites*
Distribution: Reef flat, slopes and lagoon
Characteristics: Common. *Porites* is characterized by the smallness of its corallites (2 millimetres across), a feature which gives colonies their smooth appearance, and by the compactness of its skeleton. *Porites* comprises many species but the two most common growth forms are branching colonies and rounded heads. Some of the heads can only be identified to particular species from laboratory examination. *Porites* forms some of the largest colonies found on a reef. The size of these colonies varies from a few centimetres to over 5 metres (16 ft) across. Common colours are green, grey, brown, pink and purple.

Left: Large growth form, Heron Island
Below: Growth form in lagoon, Heron Island
Bottom: Typical growth form, Lizard Island
Right: Branching form, Lizard Island

Family: FAVIIDAE
Genus: *Favia*
Distribution: Reef flat, slopes and lagoon
Characteristics: Common. Species of this coral occur as rounded colonies in which the irregularly shaped corallites are always separated. Some have evenly rounded corallites. The colour is normally brown or green. An average size is 20 centimetres (8 in.) across.
Below: Heron Island, photo: Jean Deas; Left: Lizard Island; Top right: Close-up, Heron Island; Bottom right: Lizard Island

Family: FAVIIDAE
Genus: *Favites*
Common name: Honey-Comb Coral
Distribution: Reef flat, slopes and lagoon
Characteristics: Common. Similar to *Favia*, except the corallites are joined together at their edges. The corallites are often polygonal in shape. Colours are brown, green, or brown with green centres. Size similar to *Favia*.

Left: Lizard Island; Below: Lizard Island; Top right: Heron Island, photo: Jean Deas; Bottom right: Palfrey Island

Family: FAVIIDAE
Genus: *Oulophyllia*
Distribution: Reef slopes
Characteristics: Uncommon. There is a slight resemblance between *Oulophyllia* and the brain corals (*Platygyra* and *Leptoria*) as both occur with long sinuous valleys. However, in *Oulophyllia* these valleys are much deeper and broader. *Oulophyllia* usually occurs as rounded colonies, of a lime-green or brown colour. An average colony could be 25 centimetres (10 in.) across.
Heron Island

Family: FAVIIDAE
Genus: *Goniastrea*
Distribution: Reef flat, slopes and lagoon
Characteristics: Common. This coral grows as round colonies. In some species the corallites tend to be meandroid, but the sinuous grooves are not as long as in the brain corals. In other species the corallites resemble those of *Favites,* but are smaller and finer. *Goniastrea* can be distinguished from *Platygyra* by the presence of separate corallite mouths along the grooves. Colours are brown, green or grey and an average colony would be 15 centimetres (6 in.) across. However some species grow larger.
Heron Island

Family: FAVIIDAE
Genus: *Platygyra*
Common name: Brain Coral
Distribution: Reef flat, slopes and lagoon
Characteristics: Common. Very similar to *Leptoria* and can only be separated by comparing skeletons, as *Platygyra* lacks the continuous columella. It also tends to have a more uneven appearance. Colours are green, brown and greyish. The size is similar to *Leptoria*.
Below: Heron Island; Left: close-up, Heron Island

Family: FAVIIDAE
Genus: *Leptoria*
Common name: Brain Coral
Distribution: Reef flat, slopes and lagoon
Characteristics: Common. This coral and the closely related *Platygyra* are the so-called 'Brain Corals' of the Great Barrier Reef. The usual growth form is a rounded colony which can be up to 2 metres (6 ft) across, but is commonly smaller. The corallites are highly meandroid with long sinuous valleys. In a bleached skeleton they show a continuous ridge (columella) at the bottom of each valley. Colours can be green or brown.
Below: Wistari Reef; Right: Heron Island; Top right: Heron Island

Family: FAVIIDAE
Genus: *Hydnophora*
Distribution: Reef flat, slopes and lagoon
Characteristics: Uncommon. The surface of *Hydnophora* consists of many small projections, often in the form of cones, making this coral easy to recognize. Species of *Hydnophora* occur as either branching, rounded, or encrusting colonies. The colour is normally brown, grey or green, and an average colony might be 5 centimetres (2 in.) across.
Left: Lizard Island; Below: Lizard Island

Family: FAVIIDAE
Genus: *Montastrea*
Distribution: Reef flat
Characteristics: Uncommon. This coral is similar to *Favia* in that its corallites are separate from each other, but in *Montastrea* the corallites have a circular shape and are of different sizes. *Montastrea* forms round colonies that are normally brown in colour, and average 10 centimetres (4 in.) across.
Right: Lizard Island; Below: Lizard Island

Family: FAVIIDAE
Genus: *Leptastrea*
Common name: Honey-Comb Coral
Distribution: Reef flat
Characteristics: Uncommon. Resembles a small and fine *Favites*, and usually occurs as round colonies. The colour is normally brown or a light purple.
Lizard Island

Family: FAVIIDAE
Genus: *Cyphastrea*
Distribution: Reef flat and reef front
Characteristics: Uncommon. This coral resembles a small *Montastrea*. *Cyphastrea* colonies are usually round in shape and seldom exceed 20 centimetres (8 in.) in diameter. The individual corallites are easily discernible and usually well separated. The colour is most often shades of brown.
Lizard Island

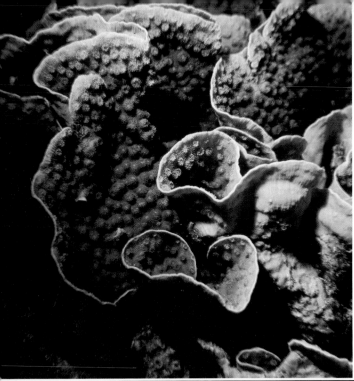

Family: FAVIIDAE
Genus: *Echinopora*
Distribution: Reef slopes
Characteristics: Common. This coral either occurs as branching colonies or as thin folia or sheets. On all species the corallites are numerous, large, and separated by ridged or spiky areas. The colours are usually green or brown and a colony of average size might be 20 centimetres (8 in.) across.
Below: Lizard Island; Left: Palfrey Island

Family: FAVIIDAE
Genus: *Trachyphyllia*
Distribution: Off-reef floor on sand bottom near reefs
Characteristics: Uncommon. This coral is found on sandy or even muddy bottoms most commonly on the off-reef floor on the more protected leeward areas of reefs. It lies free and occurs almost buried and may therefore be inconspicuous. *Trachyphyllia* is a colonial coral composed of a small number of polyps, and the average size of a colony might be 10 centimetres (4 in.) in diameter. The corallites tend to have a very irregular shape. Common colours are green, brown, reddish and grey.
Right: On the sandy bottom, Lizard Island; Below: Close-up, Lizard Island

Family: OCULINIDAE
Genus: *Galaxea*
Distribution: Reef flat and upper reef slope
Characteristics: Uncommon. This coral is easy to recognize because its corallites are well separated from each other and have long spiky septa giving it a very prickly feel. Colonies are low and often round in shape. An average colony will be 20 centimetres (8 in.) across and the colours are normally green or brown.
Below: Heron Island; Left: Lizard Island

Family: OCULINIDAE
Genus: *Acrhelia*
Distribution: Reef flats and slopes
Characteristics: Uncommon. *Acrhelia* resembles a
branching form of the related *Galaxea* and has a
fragile and beautiful skeleton. It grows as low
bushes and has a prickly appearance. It is usually
grey or brown in colour and colonies can average
15 centimetres (6 in.) across.
Lizard Island

Family: MERULINIDAE
Genus: *Merulina*
Distribution: Reef flat and slopes
Characteristics: Uncommon. This coral grows in the form of irregular sheets or plates covered by ridges that tend to radiate outwards. The central area of each colony is often covered by raised growths. The colour is generally brown or lavender. An average colony will be about 30 centimetres (12 in.) across.
Heron Island

Family: MERULINIDAE
Genus: *Clavarina*
Distribution: Lagoon and protected reef slopes
Characteristics: Uncommon. A thin branching, rather fragile coral closely related to *Merulina*. The colour is usually grey, brown or off-white and colonies often grow to 1 metre (3 ft) or more in height, with individual branches seldom reaching 1 centimetre in thickness.
Lizard Island

Family: MERULINIDAE
Genus: *Scapophyllia*
Distribution: Reef slopes
Characteristics: Uncommon. This coral resembles a massive and rounded *Merulina*. An average size might be 10 centimetres (4 in.) across and the usual colours are brown or grey.
Lizard Island

Family: MUSSIDAE
Genus: *Lobophyllia*
Characteristics: Common. This coral is characterized by the large size of its corallites which average 3 centimetres (1 in.) across. These are quite separate from each other, circular or irregular in shape. In some forms, a long stalk supports each one. Although they appear fleshy, the corallites are prickly to the touch, because of spikes on the underlying skeleton. The colour is generally green, brown or occasionally red. Colonies vary in size from 10 centimetres (4 in.) to over 1 metre (3 ft) across.
Below: Heron Island, photo: Jean Deas; Bottom: A dead corallite amid living ones, Lizard Island; Right: Heron Island

Family: MUSSIDAE
Genus: *Symphyllia*
Distribution: Reef flat and slopes
Characteristics: Common. This coral is very similar to the closely related *Lobophyllia* except that in *Symphyllia* all walls of the adjacent corallite series are joined together and form a massive structure.
Left: Heron Island, Below: Lizard Island

Family: MUSSIDAE
Genera: *Homophyllia, Parascolymia, Acanthophyllia*
Common name: Solitary Mussids
Distribution: Reef slopes
Characteristics: Common. These corals are grouped together, but, in fact, comprise several genera, *Homophyllia, Parascolymia* and *Acanthophyllia*. The solitary mussids consist of a single polyp rather like that of *Lobophyllia*, to which they are related. They are often found on cliff faces and have colours that commonly fluoresce under ultraviolet light. Common colours are red, brown, grey and green. The corals average 5 centimetres (2 in.) across.

Below: Heron Island; Right: Lizard Island, photo: Jean Deas

Family: PECTINIIDAE
Genus: *Echinophyllia*
Distribution: Reef slopes
Characteristics: Uncommon. This coral normally grows as irregular encrusting plates having a very uneven surface. The corallites all point upwards. An average colony might be 20 centimetres (8 in.) across. The colour is green or brown.
Lizard Island

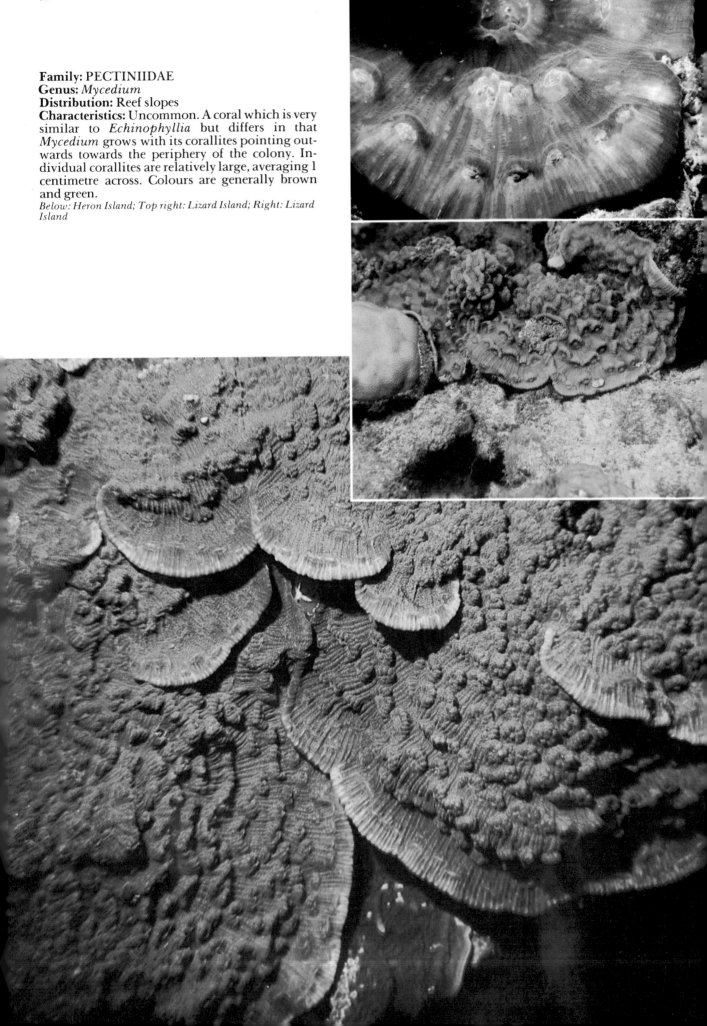

Family: PECTINIIDAE
Genus: *Mycedium*
Distribution: Reef slopes
Characteristics: Uncommon. A coral which is very similar to *Echinophyllia* but differs in that *Mycedium* grows with its corallites pointing outwards towards the periphery of the colony. Individual corallites are relatively large, averaging 1 centimetre across. Colours are generally brown and green.
Below: Heron Island; Top right: Lizard Island; Right: Lizard Island

Family: PECTINIIDAE
Genus: *Pectinia*
Common name: Carnation Coral
Distribution: Reef slopes and lagoon
Characteristics: Uncommon. This coral forms colonies which have a convoluted or leafy appearance. They can be very fragile. The colour is generally brown and the size ranges from 10-40 centimetres (4-16 in.) across.
Lizard Island

Family: CARYOPHYLLIIDAE
Genus: *Euphyllia*
Distribution: Reef slopes and lagoon
Characteristics: Uncommon. The skeleton of this coral is usually obscured by its extended bubble-like tentacles. They retract when touched, revealing the prominent septa. Species of *Euphyllia* occur as small isolated colonies or larger aggregations. The colour is habitually grey and the size varies from 5 centimetres (2 in.) to over 1 metre (3 ft) across.
Heron Island, photo: Jean Deas

Family: DENDROPHYLLIIDAE
Genus: *Dendrophyllia*
Common name: Tree Coral
Distribution: Reef slope, caves, beneath overhangs on cliff face
Characteristics: Uncommon. This coral is generally easy to recognize due to its rather tree-like growth form. It consists of 1 main stem with 'branches' growing out from either side. Colonies can be up to 50 centimetres (20 in.) long or, in some species, only 4-5 centimetres (2 in.) and stems average about 1 centimetre thick. Common colours are shades of green, black, orange and brown. The long yellow tentacles of the expanded polyp resemble flowers.
Below: Lizard Island; Left: Close-up, Lizard Island; Top right: Close-up, Heron Island; Bottom right: Close-up, Heron Island, photo: Jean Deas

Family: DENDROPHYLLIIDAE
Genus: *Tubastrea*
Distribution: Reef slopes, caves and underneath overhangs
Characteristics: Uncommon. This ahermatypic coral is similar to the closely related *Dendrophyllia*, but differs in that it does not show the tendency to branch. The single tube-shaped corallites all grow outwards from a common base. Like *Dendrophyllia*, the extended tentacles are long, and resemble flowers. Colours are usually orange, yellow or green and the size of an average colony would be 4 centimetres (1½ in.) across.
Left: A typical location of the colonies under an overhang, Heron Island
Bottom:The polyps retracted and out, Heron Island
Below: An extreme close-up, Heron Island, photo: Jean Deas

Family: DENDROPHYLLIIDAE
Genus: *Heteropsammia*
Common Name: Button Coral
Distribution: Sandy bottoms near reefs at depths usually greater than 10 metres (33 ft)
Characteristics: Uncommon. Its form is distinctive. An interesting feature of this coral is its relationship with a marine worm which lives within the base of the coral and is able to move the coral slowly about the bottom. The colours can be green, brown or yellow and average size is 2 centimetres across.
Lizard Island

Family: DENDROPHYLLIIDAE
Genus: *Turbinaria*
Common name: Vase Coral
Distribution: Lower reef slopes and lagoon
Characteristics: Common. With one exception, a semi-branching form, species of this coral occur as sheets or folia. Colonies are usually convoluted or have a vase-like shape. The corallites are relatively large, protuberant and separated by smooth areas. Common colours are light grey or green, and the size varies from 5 centimetres (2 in.) to 1 metre (3 ft) across.
Below: Lizard Island; Left: Lizard Island; Right: Heron Island; Page 84: Heron Island

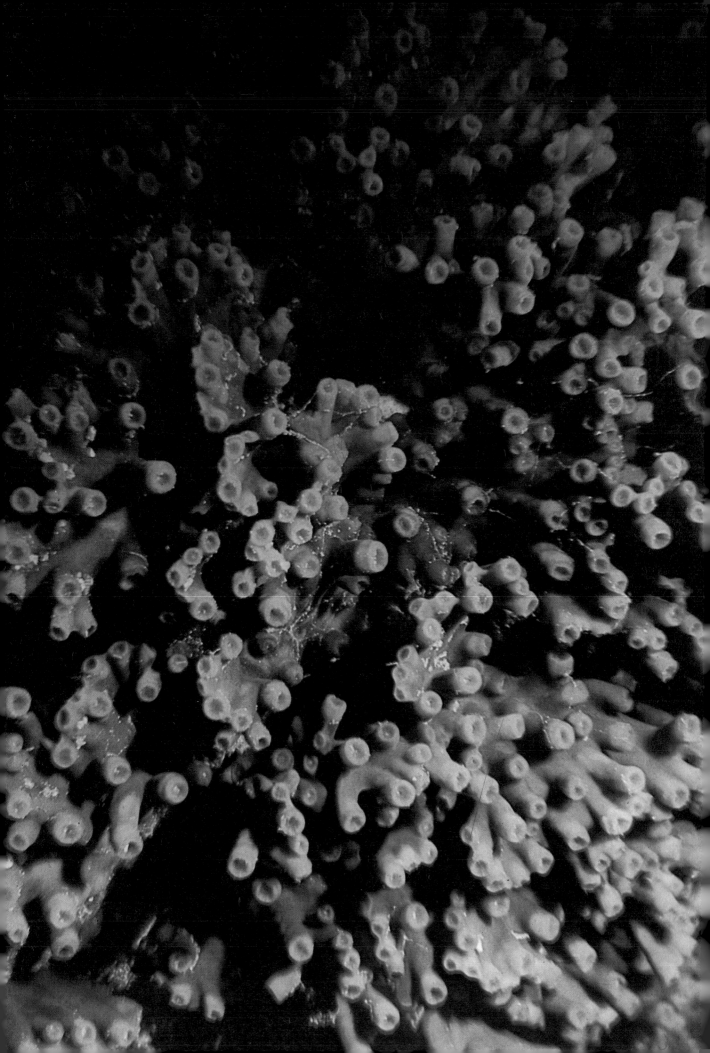

85

Hydrozoa

Order: MILLEPORINA
Family: MILLEPORIDAE
Genus: *Millepora*
Common name: Stinging or Fire Coral
Distribution: Reef flat, slopes and lagoon
Characteristics: Common. This hydrozoan coral can usually be recognized by its smooth surface and form. *Millepora,* comprising several species, occurs in a variety of growth forms: as branching colonies, encrusting growths, and as colonies with irregular vertical plates. Like other hydrozoans, this coral is capable of giving a nettle-like sting. A common colour is brown with yellow tips. The size of the colonies is variable and can be up to several metres across.

Below left: Lizard Island; Below right: Heron Island; Bottom left: Close-up, Heron Island, photo: Jean Deas; Bottom right: Lizard Island

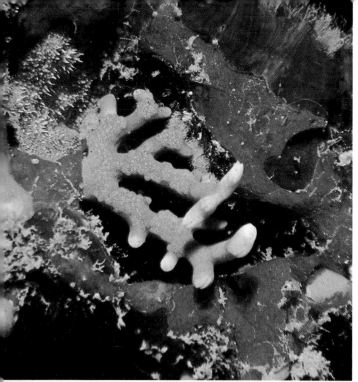

Order: STYLASTERINA
Family: STYLASTERIDAE
Genus: *Distichopora*
Distribution: Under ledges and in caves on reef slopes
Characteristics: Uncommon. This small and inconspicuous coral is often difficult to find. *Distichopora* grows as a low series of branches, having a smooth appearance and tending to be flattened in 1 plane. Colours are brown, orange or purple, with the branches often having whitish tips. An average colony would be 5 centimetres (2 in.) high.

Left: Close-up, Heron Island; Below: Close-up, Lizard Island

Order: STYLASTERINA
Family: STYLASTERIDAE
Genus: *Stylaster*
Distribution: Under ledges, in caves on reef slopes
Characteristics: Uncommon. *Stylaster* resembles
Distichopora except that it tends to grow as a more
complex series of branches. The colony generally
grows in 1 plane, and the most common colours
are pink and orange.
*Below: close-up, Heron Island, photo: Jean Deas; Page 88:
Close-up, Lizard Island*

Octocorallia

STOLONIFERA

The stoloniferans are primitive alcyonarians. They are small, always fixed, and have wedge- or horn-shaped bodies without an axial skeleton. They can form colonies, in which the single polyps are joined to each other by means of supine, cord-like stolons.

Order: STOLONIFERA
Family: CLAVULARIIDAE
Genus: *Clavularia*
Heron Island

Order: STOLONIFERA
Family: TUBIPORIDAE
Scientific name: *Tubipora musica*
Common name: Organ-pipe Coral
Distribution: Reef flat, slopes and lagoon
Characteristics: Common. The skeleton of this coral is easy to recognize because of its red colour, and form which consists of parallel tubes. On the reef, the grey-green polyps, which are always extended, cover the skeleton. Most colonies are round and can average 10 centimetres (4 in.) across.

Below: Colony, Heron Island; Left: Close-up, showing polyps emerging after being touched, Heron Island

TELESTACEA

The species can be recognized by their delicate colonies which can form monopodial, sympodial, or dense branching forms. The pinnate lateral polyps can often be seen open during daylight, especially if there is a current flowing. The branches are often covered in a red or orange sponge.

Order: TELESTACEA
Family: TELESTIDAE
Genus: *Telesto*
Common name: Soft Coral
Distribution: Low-tide level down to great depths

Below: Typical colony, this one is overgrown with a red sponge, Heron Island
Right: Close-up of the same colony, Heron Island

ALCYONACEA

Alcyonarians are all colonial growths, usually very colourful and, at times, luminescent. The skeleton is composed of separate or fused spicules. The formation growths are of a mushroom or tuber-shape and of a fleshy or hard consistency. The polyps with their eight feathery tentacles are embedded in the connecting body mass which is supported by the calcareous sclerites. These tiny spicules which give the firmness in the colony are among the most prominent characteristics used in the classification of this division. They can be found on the sea bed, on coral reefs, rocks or sandy or muddy ground on stones. They range from small fleshy lobes to giants, about 2 metres (6 ft) long. Colours can be red, orange, purple, white, lavender and various shades in between.

Order: ALCYONACEA
Family: NIDALIIDAE
Genus: *Siphonogorgia*
Common name: Soft Coral
Lizard Island

Order: ALCYONACEA
Family: ALCYONIIDAE
Genus: *Sinularia*
Common name: Soft Coral
Below: Heron Island; Right: Sykes Reef

Order: ALCYONACEA
Family: ALCYONIIDAE
Genus: *Cladiella*
Common name: Soft Coral
Heron Island

95

Order: ALCYONACEA
Family: ALCYONIIDAE
Scientific name: *Sarcophyton trocheliophorum*
Common name: Soft Coral
Distribution: Reef flat and reef front
Characteristics: Common. A soft coral easily recognized by its fleshy, plant-like appearance. It can often be seen with its polyps expanded during the daytime. However, when they contract the surface becomes very smooth. Colours vary between green and brown. An individual colony can be up to 1 metre (3 ft) across. Many colonies can cover large areas of reef.
Below: Colony, Heron Island; Right: Close-up of the polyps, Heron Island

Order: ALCYONACEA
Family: XENIIDAE
Scientific name: *Xenia elongata*
Common name: Soft Coral
Distribution: Reef flat and lagoon
Characteristics: Common. This soft coral is easily recognized and grows in small clumps most commonly on the reef flat. A characteristic feature of *Xenia* is the large pinnate tentacles which are often observed opening and closing. An average colony can be 10 centimetres (4 in.) across and the colour is normally grey or light brown.
Left: Colony, Lizard Island; Below: Close-up of the polyps, Heron Island

Order: ALCYONACEA
Family: NEPHTHEIDAE
Genus: *Litophyton*
Common name: Soft Coral
Heron Island

Order: ALCYONACEA
Family: NEPHTHEIDAE
Genus: *Dendronephthya*
Common name: Soft Coral
Left: Heron Island; Below: Lizard Island

GORGONACEA

The gorgonians or horny corals form colonies with an axial skeleton. They are mostly branched and attached with a short main trunk to the sea bed, vertical walls or underneath ledges and in caves. The skeleton is made of a horny substance known as gorgonin. It is surrounded by a softer sheath in which the polyps are embedded. Of particular beauty are the fan-like forms. Their numerous short branches are interwoven and fused together to convert the colony into a lattice of red, orange, yellow or purple 'lace'. They are usually found at moderate depths—most often, perhaps around 18-30 metres (60-100 ft)—in places in some shade from the strongest light and where they are not too exposed to waves and currents. Gorgonians, however, do occur at all depths and the larger deep water colonies may reach 2.5 metres (8 ft) or more. The gorgonians with their interlaced branches are a standing in-

vitation to settling larvae and become the hosts of sponges, hydroids, feather stars, shells, bryozoans, crustaceans and worms. At times these stimulate the host to pathological growth as it adjusts to its guests.

Order: GORGONACEA
Family: ELLISELLIDAE
Common name: Sea Whip
Bottom: Lizard Island; Below: Close-up, Lizard Island

Order: GORGONACEA
Family: ELLISELLIDAE
Genus: *Junceella*
Common name: Sea Whip
Lizard Island

Order: GORGONACEA
Family: SUBERGORGIIDAE
Genus: *Subergorgia*
Common name: Sea Fan
Heron Island

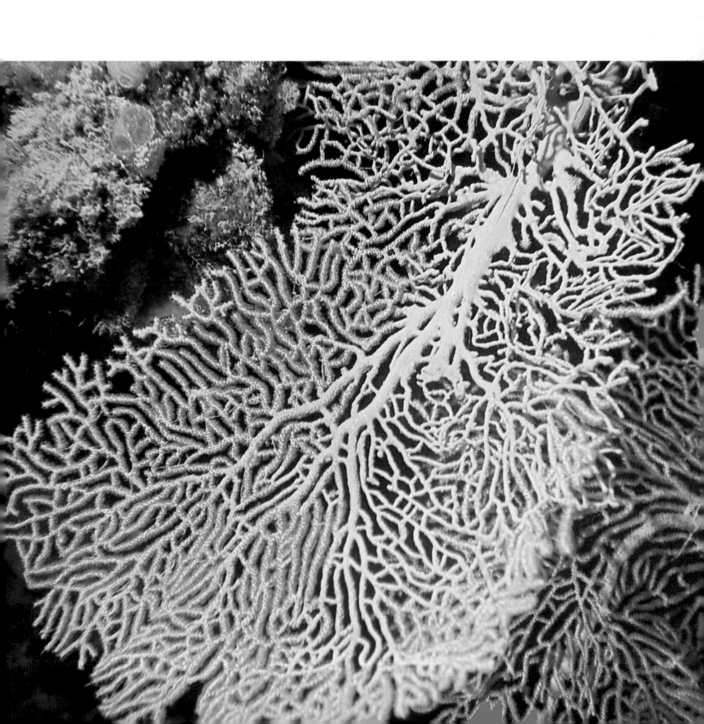

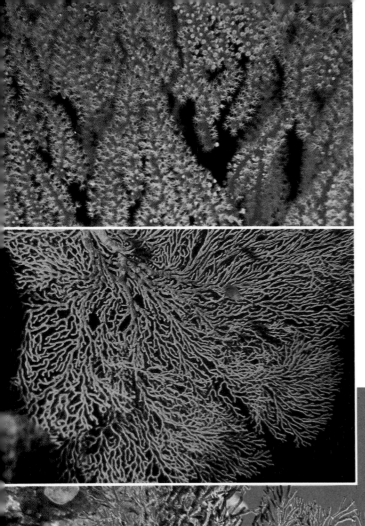

Order: GORGONACEA
Family: MELITHAEIDAE
Genus: *Melithaea*
Common name: Sea Fan
*Left: Lizard Island; Below: Heron Island; Top left: Close-up,
Heron Island*

Order: GORGONACEA
Family: PARAMURICEIDAE
Genus: *Paramuricea*
Common name: Sea Fan
Below: Close-up, Lizard Island; Right: Close-up, Lizard Island

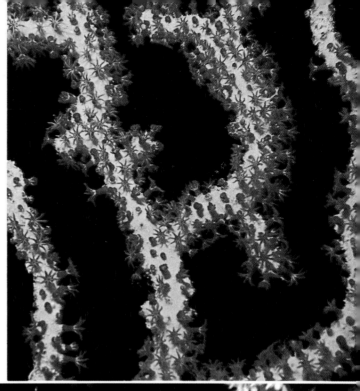

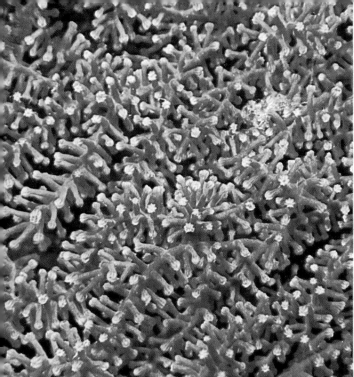

Order: GORGONACEA
Family: ACANTHOGORGIIDAE
Common name: Sea Fan
Below: Close-up, Heron Island; Left: Close-up, Lizard Island

105

Order: GORGONACEA
Family: PLEXAURIDAE
Genus: *Rumphella*
Heron Island

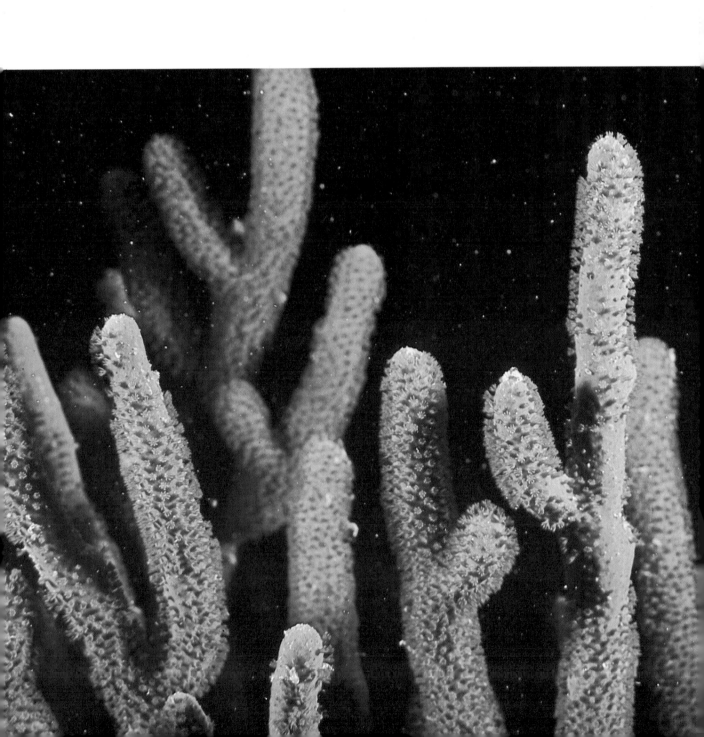

PENNATULACEA

The sea-pens are half-sedentary coral colonies with an axial skeleton, always without branches. They were named sea-pens in the days when a pen was still the feathered quill of a bird. Most species are beautiful shades of red, yellow or purple. The fleshy, cylindrical, whip or feather-shaped bodies are anchored in the soft substratum by means of an expansible bulbous stolon which bears no polyps. The upper part with the polyps is called the polypar. They range in size from a few centimetres to over 1 metre (3 ft) in height. The colonies usually contract sharply when disturbed, and they are noted for their bright luminescence, usually blue or violet, but sometimes greenish or yellowish.

Order: PENNATULACEA
Family: PENNATULIDAE
Genus: *Pennatula*
Common name: Sea-pen
Lizard Island

COENOTHECALIA

The blue coral of the tropical Indo-Pacific shores
(*Heliopora*) belongs to this order; it has a massive
blue calcareous skeleton.

Order: COENOTHECALIA
Family: HELIOPORIDAE
Scientific name: *Heliopora coerulea*
Common name: Blue Coral
Distribution: Reef flat, lagoon and reef front
Characteristics: This coral grows in low clumps,
having a smooth surface. It is only found in the
northern waters of the Great Barrier Reef,
northern Australia and northern Western
Australia. The skeleton is a light blue colour but
this is obscured by the tissues which are usually
grey or brown in the living coral. An average
colony might be 20 centimetres (8 in.) across, but
colonies over 1 metre (3 ft) across are not uncommon.
*Below: Colony, Lizard Island, photo: Gordon Cox; Page 108:
Close-up, Lizard Island, photo: Gordon Cox*

ANTIPATHARIA

The Antipatharia are not Octocorals, but are commonly grouped as a special sub-class.

The antipatharians are slender whip-like or branching colonies ranging from a few centimetres to over a metre (several feet) high. They often resemble the sea fans in their plant-like form with the main basal stalk attached to the substrate. The small polyps normally have only 6 tentacles. They usually inhabit the deep waters. The thicker, more solid, skeleton of the black coral is often cut and polished into jewellery.

Order: ANTIPATHARIA
Family: ANTIPATHIDAE
Genus: *Antipathes*
Right: Lizard Island; Below: Close-up, Lizard Island

Order: ANTIPATHARIA
Family: ANTIPATHIDAE
Scientific name: *Cirrhipathes anguinus*
Common name: Whip Coral
Characteristics: The sea whip consists of long whip-like branches usually found growing out from the sides of coral drop-offs, the surfaces of which are covered with small, barely visible polyps. It was only a few years ago that the association between the whip corals and the small goby, *Cottogobius yongei*, was discovered. Sometimes one can locate 2 fish on a whip, possibly male and female. The fishes can adopt the colour of their host. The fishes have also been found on the gorgonian, *Junceella* sp.

Below: Lizard Island; Left: The tiny goby, approximately 2-3 centimetres (1 in.) on a sea whip

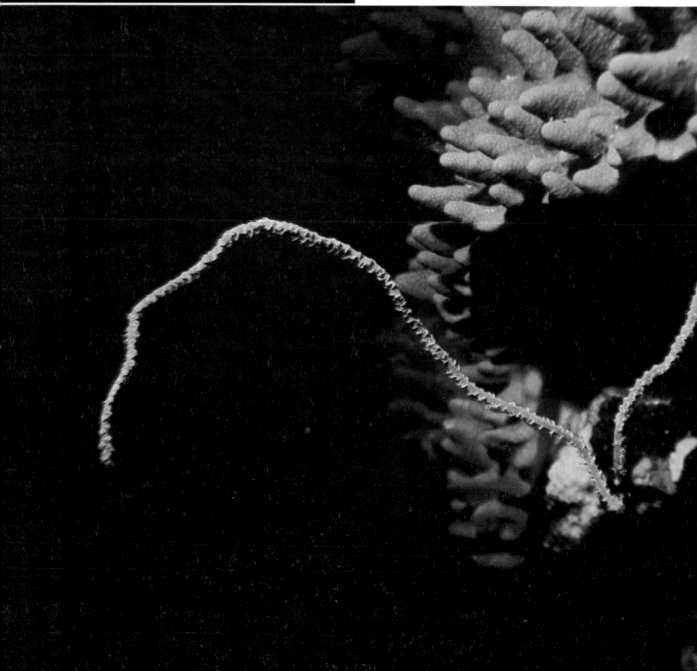

Section II Coral Reefs of The Great Barrier Reef

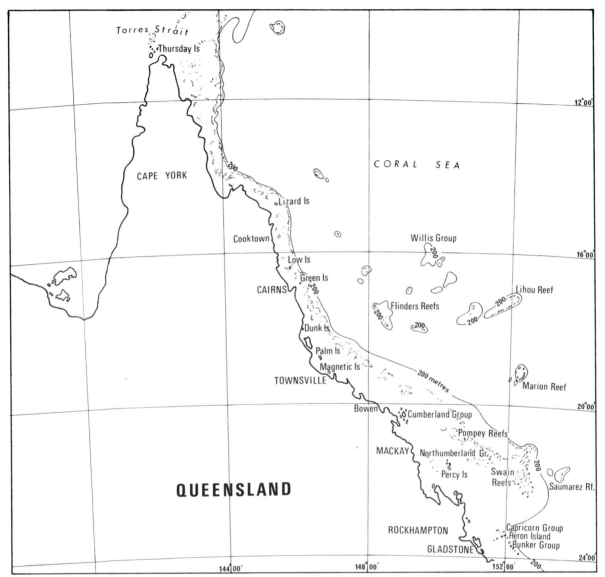

Fig. 5. Australia's Great Barrier Reef extends from latitude 24°S in the south to 10°S in the north

It has been estimated that the Great Barrier Reef comprises over 2500 individual reefs. The Great Barrier Reef is in fact a very large coral reef province approximately 2000 kilometres (1200 miles) long, and as such is one of the largest in the world. It is located on the Queensland continental shelf, although it shares its flora and fauna with other coral reefs of the Indian and Pacific Oceans.

The Great Barrier Reef has been subdivided by W.G.H. Maxwell in his *Atlas of the Great Barrier Reef* into three major regions.

1. The Northern Region: from latitude 10°S to 16°S
2. The Central Region: from latitude 16°S to 21°S
3. The Southern Region: from latitude 21°S to 24°S

This division is based on bathymetry (depth of water) which deepens as one proceeds south and also on changes in reef type. It provides a useful reference for detailed studies and comparisons of reefs within the Great Barrier Reef province.

1. The Formation of Coral Reefs

a. What is a Reef?

The Concise Oxford Dictionary defines a reef as a ridge of rock or shingle or sand at or just above or below the surface of the water. This definition includes the two major features of coral reefs; the first is they are composed of rock in the form of limestone, and the second feature is they never extend very far above sea level. The tops of coral reefs are only exposed to the air for a short time during low tides.

Coral reefs, as the name suggests, have corals living on their surface, but it is important to realize that corals are not the only organisms growing on coral reefs. It is also important to understand that the major part of a coral reef is non-living limestone and the living plants and animals only form a thin veneer on its surface. Coral reefs occur in many different shapes—and vary in size from a few metres across in lagoon reefs to over 10 kilometres (6 miles) long in some wall or platform reefs.

Coral reefs consist of consolidated limestone debris, and, once established, tend to 'grow' both upwards and outwards when the plants and animals living on their surface are able to flourish. The life and death processes of these organisms result in a steady contribution of their skeletal remains to the structure of the reef. This is especially significant in tropic seas where plants and animals tend to grow rapidly. Furthermore, tropic seas contain large amounts of dissolved limestone which is used by 'reef-forming' organisms to construct their shells and skeletons which upon death accumulate to form reefs.

b. The Role of Corals in Reef Formation

Corals constitute the major 'building blocks' in reefs and their importance is dependent upon the coral-zooxanthellae association which enables corals to produce limestone in a very efficient manner. However, corals are not always the dominant benthic (bottom-dwelling) organism on reefs. When reef rock is closely examined it is found that coral skeletons are not the only component. Together with the larger molluscs, they provide the coarser individual pieces of debris scattered through a matrix of sand-sized fragments of calcareous (limestone) plants, Foraminifera, and other organisms. It is interesting to note that coral reefs may possess large areas that are almost devoid of living corals due to unfavourable living conditions. Nevertheless, on most reefs of the Great Barrier Reef, the living corals are the most obvious animal, with the genus *Acropora* often the most abundant.

Coral skeletons make up varying percentages of reef rock. These skeletons range from the fast-growing *Acropora* whose 'branches' may average 1 centimetre in diameter to massive *Porites*, where a single colony may be several metres across. All Scleractinia and the limestone secreting species of hydrozoans contribute to the formation of reefs. The skeletons of dead corals are transported and accumulated by water movement to be cemented into the structure of the reef by calcareous plants.

c. The Role of Plants in Reef Formation

Algae are marine plants and their role in the formation and maintenance of coral reefs is almost as important as and, in places, possibly more so than the corals themselves. There are many species of algae found on coral reefs, but the most important in reef formation are the calcareous algae. The calcareous secreting algae comprise numerous genera and species (e.g. *Lithophyllum* and *Lithothamnium*), all having the ability to produce limestone in the form of an encrustation over the reef, which acts as a cement, binding together loose fragments of debris to form a hard tough surface. This surface is especially characteristic of the windward areas of a reef where wave action is most violent; here the algae give the reef surface a smooth pavement-like texture of a pink, purple or light brown colour. If coral skeletons could be considered the 'bricks' that make up a reef, the calcareous algae could be likened to the 'cement' that holds these together.

Another marine plant that is important in the formation of coral reefs because it produces large amounts of sand-sized calcareous fragments is *Halimeda*. This genus contains numerous species which produce limestone as a thin outer layer over their fleshy leaf-like segments. This limestone becomes incorporated into the structure of the reef upon the death of the plant and, in some areas, sand made up primarily of *Halimeda* fragments is very common.

d. The Origin of the Great Barrier Reef

In considering the origin of the Great Barrier Reef let us first describe how any reef is formed and then we will consider the conditions peculiar to the Great Barrier Reef itself.

A reef begins to form when a shallow area of the sea floor within the tropics can provide suitable conditions to encourage and support colonization by benthic plants and animals. Such conditions include the presence of a solid surface that is free from excessive mud and sand. This must be in relatively clear, warm, sea water having sufficient movement to circulate oxygen and plankton amongst the colonizing organisms. As these plants and animals grow ever upwards and outwards they eventually die and their skeletons and shells accumulate and are cemented together by the calcareous algae. A limestone mound is formed which represents the beginnings of a reef. Corals, molluscs, and algae flourish on the surface of the newly formed reef and, when they die, add to its bulk. This process, over thousands of years, raises the reef towards the surface of the water, and it could be described as a 'growing' reef. When this reef finally reaches the surface, upward growth is checked, but lateral (sideways) growth continues. Different types of reefs form in different ways; for example a fringing reef will grow outwards from a continental island, while a platform reef may form over a solid irregularity anywhere on a continental shelf.

The reefs of Australia's Great Barrier Reef probably originated in this way. However, special conditions favoured their development and continued growth. The first of these conditions was the occurrence of a broad continental shelf off the coast of Queensland which provided a suitable base for coral colonization in shallow water. A second factor that initiated reef development in the past, but was not peculiar to this region alone, was the fluctuation in sea level that occurred in the Pleistocene era many thousands of years ago. During this time the sea level was on several occasions much lower than its present level and parts of the continental shelf were exposed to air.

When the sea began to rise, taking thousands of years, benthic organisms colonizing suitable areas were able to keep pace with the rising sea level until the present reefs were formed. Warm water and favourable oceanic and tidally induced currents maintain flourishing coral growths throughout the Great Barrier Reef at the present time.

2. Types of Coral Reefs

There are many different types of reefs found within the Great Barrier Reef province and it is useful to be able to identify them so work and observations on one can be compared to another. Coral reefs have been divided into two main categories. The first comprise the oceanic reefs which are found in the open ocean, have a non-limestone reef base and occur in deep water (over

Wistari Reef Lagoon, showing a complex pattern of lagoonal reefs.

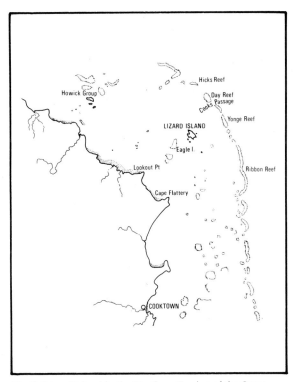

100 fathoms). The second group includes the shelf reefs which occur in relatively shallow water on continental shelves. Since the Great Barrier Reef occurs on the Queensland shelf we will confine our attention to the shelf reefs. Fringing reefs are common to both groups although they are best developed in the oceanic province.

A recent classification of reefs, dealing specifically with those of the Great Barrier Reef, is included here. This classification is after that in the *Atlas of the Great Barrier Reef* by W.G.H. Maxwell. Although it isn't important at this stage for the reader to memorize all the different reef types, it is important to note the complex shape of many reefs, and the fact that some of the reef types appear to have evolved from others. Reefs didn't just come into being suddenly as we see them now, but are continually growing, changing shape and sometimes slowly developing from one shape to another.

The separation of reefs into different types is based on shape, central structure (e.g. development of a lagoon), general zonation and their location on the continental shelf. Generally speaking, the reefs of the Great Barrier Reef can be separated into three major categories: wall reefs or linear reefs (commonly referred to as outer barrier

Fig. 6. Lizard Island in the Northern Region of the Great Barrier Reef is the site of a small research station. Within a radius of 25 kilometres (15 miles) from Lizard Island are found most of the major reef and island types that occur in the entire Great Barrier Reef province.

Fig. 7. This classification (after Maxwell) shows the two major types of coral reefs found in the world. Shelf reefs are the most common types found on the Great Barrier Reef. This diagram illustrates their often complex shapes and the fact that many appear to be interrelated.

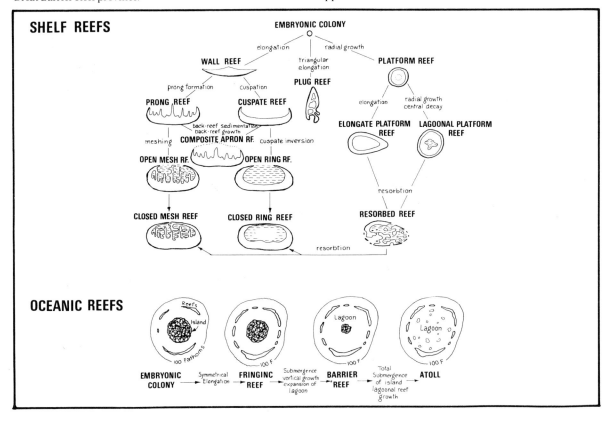

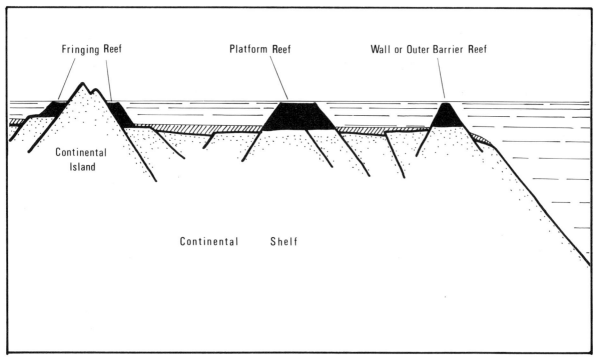

Fringing Reef   Platform Reef   Wall or Outer Barrier Reef

Continental
Island

Continental   Shelf

Fig. 8. Diagrammatic sketch showing the continental shelf and the three major types of reefs found on the Great Barrier Reef

reefs) occurring on the northern part of the province near the seaward margin of the shelf; platform reefs, which comprise the majority of reefs, have an oval shape and lie between the seaward margin of the continental shelf and the mainland; and fringing reefs that are found growing out from the shores of continental islands and the mainland.

a. Variations in Shape

From Maxwell's classification of shelf reefs it can be seen that the reefs of Australia's Great Barrier Reef can assume highly irregular and sometimes complex shapes. The 'growth' of a reef is determined by the limestone secreting plants and animals living on its surface, and the influences on their growth, either retarding or accelerating it, determine the shape of a reef.

Until a reef reaches the surface of the sea and begins to be exposed to the air during low spring tides, the major growth probably takes place in an upward direction and the shape is more or less symmetrical. Corals, molluscs, plants and many other organisms all live and die, leaving their skeletons in situ and providing a suitable base for further generations to grow upon. This process, over thousands of years, gradually raises this mound of lithified (hardened) debris until it reaches approximately mean sea level (between high and low tides). At this stage upward growth is checked as the organisms growing on the reef surface cannot tolerate prolonged exposure to air. The reef then grows in a lateral direction. If the factors influencing reef growth were uniform on all sides, reefs would be circular in shape.

However, this is not the case as the factors influencing reef growth are not uniform and affect some parts of a reef more than others. These factors include the effects of wave action, sand movement, turbidity, temperature, bottom topography and the depth of water adjacent to the reef. The most obvious factors that induce asymmetry in the shape of a growing reef are water movement (in the form of breaking waves) and bathymetry (depth of water).

The exposed margin of a reef during an exceptionally low spring tide. (Heron Island)

Left: Wilson Island showing adjacent reef flat and upper reef slopes. Note the scattered coral formations separated by sand. Typical of a shallow leeward reef slope.

Above: Wreck Island, a typical sand cay, is composed of sand derived from the surrounding reef.

Below: This clump of luxuriantly growing corals is characteristic of the leeward areas of reefs.

Above: These low growing corals are from the lower reef slopes on the windward side of a reef. (Lizard Island)

Below: A typical shallow or upper reef slope. (Heron Island)

Above: A wave breaking on the reef margin. Reefs can be a serious hazard to shipping. (Wistari Reef)

Below: Scattered growths of coral on a sandy bottom are a feature of many lagoons. (Heron Island)

Lizard Island is a continental island and has a well-developed fringing reef, especially the windward side. Lizard Island reef contains a lagoon formed by fringing reefs extending between it and nearby South and Palfrey Islands.

Above: Looking out over the lagoon with Palfrey Island on the extreme right and South Island next to it
Top right: Lagoon, Lizard Island
Bottom right: Lizard Island from the west

The prevailing wind on the Great Barrier Reef is the south-east trade wind. This produces a surf which breaks almost continuously on the south-east (or windward) side of a reef and this results in prolific growths of corals and calcareous algae. The windward side of a reef is usually very compact and has a regular outline. On the other hand, the north-west (leeward) side of a reef experiences relatively calm water since the reef acts like a breakwater. Here sand accumulates after being generated and swept across the reef by wave action. Generally this area has a more irregular margin and the more fragile corals are able to flourish here.

· The effect of water depth on reef shape is quite simple. Reefs can only grow outwards if debris can accumulate on the reef slope, but if the depth of water is too great this debris will slide down the slope and be lost. The effect of depth is best illustrated by the wall or outer barrier reefs, e.g. Ribbon Reef near Lizard Island. These reefs are located at the edge of the continental shelf. To seaward of the reefs the shelf drops away to thousands of metres, whereas behind the reefs, on the shelf, the depth of water may average only 25 metres (82 feet). These reefs are prevented from growing seawards because of the depth and so have grown sideways, giving them a long linear shape. It is interesting to note that the long, outer barrier reefs are prevented from growing together and forming one complete reef by tidal currents. These currents attain high velocities in the openings between the reefs and the scouring which results prevents corals from flourishing and thereby closing these passages.

To conclude this section here are some definitions of terms commonly used to describe natural areas on coral reefs. Not all reefs have every one of these areas and different workers may use these terms in slightly different ways, but the following are probably the most widely accepted.

One Tree Island is a coral cay composed entirely of large
fragments of corals and reef rock

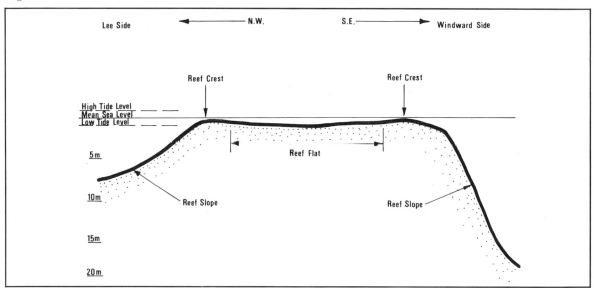

Fig. 9. Diagram of a platform reef showing major areas

Reef slope. The sloping side of a reef rising from
the sea floor to the top of a reef and having a
gradient ranging from a gentle slope to a cliff. Upper reef slope would represent shallow depth and
lower reef slope deeper water.

Reef flat. This area is essentially the entire flat top
of a reef and can be divided into the outer reef flat
and inner reef flat.

Algal rim. This area has a relatively smooth surface and is composed of calcareous algae. It is
sometimes called the reef crest since it is slightly
higher than the reef flat. The algal rim is best
developed on the windward sides of a reef where it
occupies the outer periphery of the reef flat.

Lagoon. Not all reefs contain lagoons but where
they occur they are the depressed central area of a
reef and retain water during low tide. Some
lagoons have entrances to the sea while others are
completely enclosed by the surrounding reef.
They are all relatively shallow, averaging less
than 10 metres (33 ft) in depth.

Underwater Photography as a Scientific Tool

Underwater photography is a highly technical skill requiring expensive and sophisticated equipment. It is not practical for all marine scientists to acquire this skill and to purchase equipment when top-class underwater photographers are available. The partnership of marine scientist and underwater photographer is of tremendous value to the scientist and, one hopes, to the photographer as well. This is well illustrated in the experience of the writer, and underwater photographer Walt Deas, in obtaining many of the photographs for this book. Using scuba and swimming together, the writer and photographer spent many hours beneath the sea. I, the scientist, selected the particular specimen for photography and Walt, the photographer, would photograph it

from both a distance and close up. Unencumbered by cameras, the writer was free to collect specimens for later identification. Working together in this way, the most efficient use was made of the time in the water: the writer collecting where necessary, identifying corals and recording information, with Walt photographing, using two cameras. Every roll of film Walt exposed had a high percentage of publishable photos and we very seldom had to repeat one. Many of the photographs in this book are from the photographer's collection and were taken at Heron Island, but the majority are from Lizard Island. The following section is a comment by Walt Deas on the photography.

Walter Deas photographs an *Acropora* coral at Sand Bank Reef near Lizard Island. (Photo: Jean Deas)

Steve Domm (right) with the assistance of his wife, Alison, checks a coral against his list of corals still to be photographed.

Although cameras were invented in the last century most of the vast underwater region of the world is untouched photographically. It is only in the last decade that man has begun to explore this undersea frontier. The availability of scuba diving equipment, underwater cameras and an array of suitable films has caused a remarkable growth in the field of both amateur and professional underwater photography. Many picture editors and scientific heads of departments feel it is easier to make a photographer into a diver than vice versa. But it seldom works. Most of the top underwater photographers have been diving for a minimum of ten years. A good diver-cameraman is comfortable and familiar with the alien environment and not forever worrying about the sound of his regulator, preoccupied with getting back to the surface or afraid of what may be behind him.

Taking underwater photographs is not as easy as taking them above the water. The technical problems are often seemingly absurd. Water magnifies your subject by about twenty-five per cent, light conditions are seldom the same and will vary as will the colour due to the weather, angle of sun, suspended algae and plankton, type and colour of bottom, and water depth. The photographer is working in a medium that absorbs, scatters and refracts light. The medium itself is in motion. The photographer is seldom stationary and the subject is often moving. This is a small sample of the problems encountered by the underwater photographer.

The equipment used to take the underwater photographs in this book was a Rolleiflex 3.5f camera in a Rolleimarin housing. This classic housing, designed by Han Hass, allows complete operation of all controls and very accurate focussing and viewing is possible through its prismatic viewing system. Another camera used was an electric-drive Nikon F with a 20 mm lens in a Niko-Mar I housing. This was designed by Al Giddings and is specifically designed for the Nikon F-36 motor-driven camera. Viewing is through the large prism reflex sportsfinder and every Nikon lens from 20 mm to 105 mm can be used with control of focus and aperture. Special semi-spherical ports are available for wide-angle lenses and give full angle-distortion-free photographs. A Nikonos fitted with extension tubes and 35 mm lens was used for close-ups. The Nikonos, designed in the early 1950s, is a self-contained amphibious camera. It is a small light 35 mm unit that can be fitted with various lenses and a range of accessories. I used a Metz 162 electronic flash in a custom-built perspex housing or a Pentax Super Lite II in a Sea Tite housing for lighting with the Rolleimarin. With the Niko-Mar I a powerful Sea Star III electronic flash specifically designed for underwater use was used. For close-ups with the Nikonos a low-power East Lite electronic flash, again in a custom-built perspex housing, was used. All units were fitted with E-O connectors thus making all equipment interchangeable. The surface photographs were taken with either a Hasselblad or Nikon F. The film used throughout for this book was Ektachrome X which is rated at 64 asa, however I have pushed this film to 800 asa in poor light conditions.

Index

Page numbers in italic type refer to diagrams or photographs